Managing Innovation in SMEs

Managing Innovation in SMEs: Product Development in Small Irish Electronics Firms

By

Ann Ledwith

CAMBRIDGE SCHOLARS
PUBLISHING

Managing Innovation in SMEs: Product Development in Small Irish Electronics Firms,
by Ann Ledwith

This book first published 2009

Cambridge Scholars Publishing

12 Back Chapman Street, Newcastle upon Tyne, NE6 2XX, UK

British Library Cataloguing in Publication Data
A catalogue record for this book is available from the British Library

ISBN (10): 1-4438-1375-3, ISBN (13): 978-1-4438-1375-4

TABLE OF CONTENTS

ACKNOWLEDGEMENTS

I would like to thank the following individuals and organisations for their help in the completion of this thesis. First and foremost my supervisors, Prof. John Bessant and Prof. Paul Coughlan who made the perfect supervising team and who provided me with invaluable advice and encouragement especially when completion drifted further from sight.

Thank you to all the managers with responsibility for NPD who gave up their valuable time to completed the INTERPROD questionnaires. I am especially grateful to Philip, Alan and Colin and all the staff at the three firms who participated in the case studies.

Many thanks to the researchers involved in the INTEPROD study who got me started on this project especially Prof. Bill Souder, the director of the project, and Prof Rachel Cooper who collected the UK data and provided me with access to her data set.

Thanks to the different individuals and groups at the University of Limerick who provided me with various forms of help and support. These include colleagues at the Department of Manufacturing & Operations Engineering and the Department of Management and Marketing, the SFRU, the TTI and the PhD Forum.

Finally, thank you to my family, my parents, Rita and Jimmy, brothers and sisters (especially Sarah) who bailed me out on many occasions. To Laura, Kevin, Harry, Fiona and Claire who arrived at various stages throughout this process and may some day come to appreciate it. To all the women who kept my home functioning so that I could concentrate on other less important things. And last but not least to John who was always there.

CHAPTER ONE

INTRODUCTION

1.1 Introduction

This volume examines the management of New Product Development (NPD) within small Irish firms. Much research to date into the development of new products has been based on the practices within large, industry leaders such as Rank Xerox, 3M, Intel and Eastman Kodak (see for example Bowen et al., 1994). Empirical evidence is still lacking, however, for product development in small firms (Acs & Audretsch, 1988; Berry & Taggart, 1998; Hoffman et al. 1998; March-Chordà et al. 2002)

Many authors, (Alder et al, 1989; Cooper & Kleinschmidt, 1995; Song & Parry 1997, Song & Montoya-Weiss, 1998) provide advice on what is 'best practice' in the management of NPD. However, another school of thought suggests that it is more realistic to examine which management practices are most successful in different situations (Loch, 2000; Bessant & Francis, 1997).

This study is based on the premise that certain conditions and management decisions influence the success or failure of new products. Additionally, the manner in which these conditions and decisions are managed is contingent on the size of the firm. The companies that form the focus of this research are differentiated from those usually studied on the basis of their size and their location. The impact of these two factors on the management of NPD is central to this work.

The aim of this research is to determine the effect of firm size on product development management within the Irish setting. To this end, the research will focus primarily on firms within Ireland and the UK. The management of NPD has been found to vary depending on the industry sector (Karakaya & Kobu, 1994) and on the technology level of a firm (Meyer & Roberts, 1986, Bacon et. al, 1994). To avoid the impact of these factors, this study will concentrate on firms in the electronics sector supplying to industrial customers.

This chapter is divided into four sections. The first, Section 1.2, presents the background to this project, a review of how small electronics firms in Ireland have been performing over the past few years and an indication of their importance to the Irish economy. Section 1.3 examines questions that emerge from the background information; these emerging questions are summarised to form the research question that will then guide this project. Some definitions that will be used throughout this volume are given in Section 1.4. A brief overview of the complete volume is presented in Section 1.5 and the conclusion to the chapter is in Section 1.6.

1.2 Background

The development of new products has been highlighted as a key factor in increasing competitiveness both at a company and a national level. It has been shown in the UK that innovating firms have an average market share of about five times that of non-innovators (Geroski & Machin, 1992). The link between innovation and firm growth was highlighted by Roper (1996) who found that in small Irish firms turnover growth between 1991 and 1993 was 9.7% for non-innovators but 15.7% for innovators. Additionally, Roper found that employment growth was 4.3% for non-innovators and 7.3% for innovators. Ruane & Kearns (1997) in a report based on data collected by Forfás also established a clear link between R&D activity and employment at a sectoral level.

Several studies show that product innovation is more difficult for small firms than for large firms. A study of product innovation and development in the UK, Germany and Ireland (Roper et. el., 1996), showed that small firms experienced greater difficulties in recruiting technical staff and obtaining finance for product innovation than did larger firms. Additionally, a 1994 Forfás report found that Irish-owned firms were innovating at a lower rate than their foreign-owned counterparts, and also that small firms were less likely than large firms to be involved in product innovation.

This study examines product innovation in small electronics firms in Ireland. This group of firms is defined by two characteristics; they operate within the electronics sector and they are small. There follows a brief definition of what is meant by a *small* firm and then a review of the electronics sector in Ireland highlighting where possible the performance of small electronics firms.

1.2.1 Small Firms in Ireland

The Government Task Force Report in Small Business (1994) states that 'Ireland is a nation of small businesses'. Ninety-eight percent of Ireland's estimated 160,000 businesses employ fewer than 50 people while 85% employ fewer then 10. In Ireland, small business accounts for almost a half of all employment. Within the Task Force Report small firms are defined as having fewer than 50 employees and a turnover of less than £3 million. However, this is by no means a universally accepted definition of a small firm. An example of the wide range of definitions of 'small' is given in Table 1.1. These are of particular interest to this study because they come from authors writing about innovation or NPD.

Author	Definition of small	Author	Definition of small
Walsh et al. 2002	< 100	laBahn, 1996	< 100
Nassimbeni, 2001	10 – 33	Acs & Audretsch, 1988	< 500
EU,	< 50	Ali et al., 1995	< 100
Freel, 2003	< 500	Roper, 1997	20 < 100

Table 1.1 Definitions of Small Firms

As can be seen in Table 1.1 the most common definition of small firms is that they have fewer than 100 employees. Clearly, many definitions of small firms are more complex than this, and include parameters such as turnover or number of productive units; however, the most commonly, used parameter for defining small firms is number of employees (Richardson, 1999). Therefore, for this research project a small firm will be defined as one having fewer than 100 employees.

This study is based on small electronics firms. It is useful at this point to examine the overall performance and particularly, the innovation performance of these firms in Ireland.

1.2.2 The Electronics Sector in Ireland

The information presented in this section has been collected mainly from reports published by the European Commission in 2000/2001. The data in these reports was collected between 1996 and 1999. This section is not intended to be a detailed economic analysis of the electronics sector in Ireland but rather an overview of its importance to the Irish economy during the late 1990s. The following topics will be examined:

1. The number of innovating enterprises
2. Expenditure on innovation
3. Employment levels
4. Innovation cooperation
5. Patent applications
6. Turnover due to new products
7. Exports.

The first four of these topics represent some of the inputs into the innovation process; the last three can be viewed as outputs. Within this section, data from Ireland, the UK and the EU-15 average are presented.

The industrial sectors that are examined in this study are described by NACE Rev. 1 30 – 33. Within this volume these sectors will be jointly referred to as the electronics industry. The activities included in NACE 30 – 33 are listed below:

NACE 30:	manufacture of office machinery and computers,
NACE 31:	manufacture of electric motors, generators and transformers, electricity distribution and control apparatus, as well as insulated wires and cables, accumulators, batteries and lighting equipment.
NACE 32:	manufacture of electronics components, television, radio and video equipment, as well as telephones.
NACE 33:	manufacture of medical, precision and optical instruments, watches and clocks.

There is no generally-accepted definition of the 'electronics' industry. In some EU reports NACE 29 – 33, *Machinery and equipment, electrical and optical equipment* is considered to consist of one sector. In other places, NACE 29 – 32 is defined as *Electrical machinery and electronics,* or (again) NACE 30 – 33 is considered to be *Electrical and optical equipment.* Within this study, firms classified as NACE 30 – 33 are considered to be operating within the electronics industry. The rationale behind this is that it is a reasonably broad definition, one which should encompass most firms operating in the sector. Additionally, this study concentrates on firms supplying to industrial customers and, with a few exceptions, (notably within NACE 32) the majority of firms operating within NACE 30 – 33 have industrial customers. However, NACE 29 – *Manufacture of Domestic Appliances* has been omitted as firms within this sector concentrate on domestic markets.

In 1996 the electronics industry accounted for 11.6% of the EU's total production value, or €450.7 billion. Production value within the electronics sector rose during the 10-year period 1986 – 1996. However, during this period there was also a decline in the workforce within the sector. The electronics sector has a highly-educated workforce and a higher labour productivity than the EU average. In this sector Ireland has higher than (EU) average labour productivity and low unit personnel costs.

Number of Innovation Enterprises

The second Community Innovation Survey (CIS2) conducted in 1997/98, found that Ireland had the highest number of innovating enterprises in the manufacturing sector of any of the EU-15 countries (Eurostat, 2000). For the purpose of this survey technological innovation is defined as: 'the introduction onto the market of a technologically new or significantly improved product or the implementation of a technologically-new or significantly-new process'. Seventy-three percent of manufacturing enterprises in Ireland claimed to be involved in innovation. The figures and a breakdown based on firm size are given in Table 1.2.

Firm Size	Ireland	UK	EU-15 Average
Total	73	59	51
Small (20 – 49)	68	54	44
Medium (50 – 249)	78	59	58
Large (250+)	85	81	79
NACE 29 - 33	88	70	68

Source: Statistics on Science and Technology in Europe, EC, 2000

Table 1.2 Innovating Enterprises as a % of total manufacturing enterprises - 1996

The trend seen in Table 1.2, of larger enterprises being more likely to innovate than smaller enterprises, is found in all countries throughout the EU. However, in Ireland there is a higher percentage of small manufacturing firms involved in innovation than in any other European country.

Data on the percentage of innovating firms has also been analysed across different sectors. In Ireland, the electronics sector contains the highest percentage of innovating firms of any sector in any country in Europe – 88%. Within the European countries, the group NACE 29 – 33 is also the sector with the highest percentage of innovating firms – 68%.
CIS2 also reported on the number of firms that could be classified as novel innovators. Novel innovators are defined as firms whose latest products are new not only to the firm but also to the market. Once more Ireland was found to have a very high percentage of novel innovators as is shown in Table 1.3.

Firm Size	Ireland	UK	European Average
Small (20 – 49)	23	15	16
Medium (50 – 249)	28	19	23
Large (250+)	50	37	42
NACE 30 - 33	48	32	36

Source: Statistics on Science and Technology in Europe, EC, 2000

Table 1.3 Novel Innovators as a % of total manufacturing enterprises - 1996

The electronics sector is again found to have most firms involved in novel innovation, with 36% of firms innovating. In Ireland 48% of electronics firms report that they are involved in novel innovations this is second only to the same sector in Germany.

The above data shows that innovation, including the development of new products, is a key activity for firms operating within the electronics sector. Additionally, and of importance to this study, innovation is an activity undertaken by a large number of small electronics firms.

Expenditure on Innovation

The late 1990s in Ireland was a period of unprecedented growth. The average growth rate of GDP in Ireland between 1995-1999 was in excess of 9% while the EU average was just above 2%. Table 1.4 details R&D expenditure in 1998.

	Ireland	UK	EU - 15
R&D expenditure in EU (million) and as a % of GDP	968 (1.40%)	22,865 (1.82%)	141,200 (1.86%)
Growth in R&D expenditure as a % of GDP, 1985 – 1998	71%	-18%	-6%
R&D expenditure by sector:			
Business Enterprise	74%	67%	64%
Higher Education	19%	20%	21%
Government	7%	14%	15%
R&D personnel (full time equivalent) and as a % of labour force	12,372 (1.17%)	277,500 (1.28%)	1,625,500 (1.27%)
Growth in R&D personnel as a % of labour force, 1985 - 1998	113%	-7%	8%
R&D personnel by sector[1]:			
Business Enterprise	69%	54%	56%
Higher Education	23%	36%	29%
Government	8%	10%	15%

Sources: Statistics on Science and Technology in Europe, EC, 2000, Science and Technology in Europe, Statistical pocketbook, EC, 2002

Note 1: The figures quoted in this section are compiled from both the above sources; some of the figures from the Statistical pocketbook are estimates. But cross-referencing shows that errors are less than 1% and in the context of this discussion can be ignored.

Table 1.4 R&D expenditure and employment - 1998

In 1998 Ireland's expenditure on R&D was below the EU average, though it had increased dramatically during the previous 13 years. But the very high growth of R&D expenditure in Irish firms has to be viewed in light of the very low starting position. The growth in R&D expenditure in Ireland is in contrast to the trend in the UK and Europe where R&D expenditure fell. Additionally, the breakdown of R&D expenditure in Ireland is different from that in the rest of Europe. Spending on R&D in the government sector in Ireland is below the EU average. In fact

government spending on R&D (as a percentage of total expenditure) fell from above 20% during the previous 10 years while business expenditure rose from below 50%. Much of this change can be explained by increased activity in the business sector with government expenditure failing to keep pace[1].

These changes in R&D expenditure are reflected in the number of people employed in R&D during the same time period. In terms of full time equivalents, the number of people employed in R&D in Ireland more than doubled between 1985 and 1998, from 5,341 to 12,372, however, it is still slightly below the EU average of 1.27% of the labour force in R&D.

Another measure of how much is being spent on innovation is the level of innovation expenditure as a percentage of total turnover. These data obviously relate only to business expenditure and are detailed in Table 1.5.

	Ireland	**UK**	**European Average**
Innovation Expenditure as a % of Total Turnover	3.3	3.2	3.7
Small (20 – 49)	2.8	3.3	2.5
Medium (50 – 249)	3.2	2.9	2.3
Large (250+)	3.7	3.2	4.2
NACE 29 - 33	4.9	6.1	6.3

Source: Statistics on Science and Technology in Europe, EC, 2000

Table 1.5 Innovation expenditure in manufacturing enterprises - 1996

Within the EU countries the electronics sector spends a higher percentage of turnover on innovation than any other sector. In Ireland this is not the case, innovation expenditure in the electronics sector is below the EU average for the sector.

---

[1] Government expenditure on R&D has remained relatively constant at approximately 0.32% of GDP between 1988 and 1998 (Eurostat, 2000).

The data presented in this section when viewed in contribution with the data presented in the previous section raise some questions. If so many enterprises in Ireland claim to be innovating, how is it that overall expenditure on innovation is so low? There are two possible explanations for the discrepancy between these two sets of data; either (1) Irish firms are understating their expenditure on innovation or (2) they are overstating their innovation activity. Because of tax implications it is unlikely that firms would understate their expenditure on innovation. It is more likely that the innovation activity in Irish firms is overstated; this is particularly true in large Irish firms whose innovation expenditure is more that 10% below the EU average.

Employment in High-tech sectors

Employment in the high-technology sectors is detailed in Table 1.6. The figures show that Ireland is close to the EU average in terms of the numbers employed in high-technology manufacturing industries but above the EU average for the percentage employed in electrotechnology etc. This sector includes NACE 30 – 33 and these results confirm that this sector is of importance to the Irish economy, accounting for almost 4% of total employment.

	Ireland	UK	EU 15
Total Employment[1]	1,591	27,442	145,958
Total Manufacturing[2]	291 (18%)	2,073 (18%)	31,551 (22%)
High-Tech Industries[2]:	116 (7%)	2,073 (8%)	11,916 (8%)
Chemical[3]	27 (23%)	331 (16%)	2,145 (18%)
Mechanical and Automotive[3]	30 (26%)	1078 (52%)	6,316 (53%)
Electrotech., I&C etc.[3]	59 (51%)	663 (32%)	3,456 (29%)

Source: Statistics on Science and Technology in Europe, EC, 2000
Note 1: Figures given above are in thousands. Source of total employment figures – International Labour Organisation (1999) http://laborsta.ilo.org
Note 2: Percentages given are percentages of total employment.
Note 3: Percentages given are percentages of employment in high-tech manufacturing industries.

Table 1.6 Employment in High-tech and Knowledge intensive sectors, 1999

Innovation Cooperation

As technology increases in complexity and developing new products becomes riskier and more expensive, cooperating with external organisations in R&D/innovation projects emerges as an attractive option. Table 1.7 shows the number of manufacturing enterprises that were involved in innovation cooperation in 1996.

	Ireland	**UK**	**EU - 15**
Total	36	32	27
Small (20 – 49)	26	22	19
Medium (50 – 249)	38	36	28
Large (250+)	84	53	50

Source: Statistics on Innovation in Europe, EC, 2000

Table 1.7 Percentage of innovators with innovation collaboration – 1996

Data from CIS2 show that vertical cooperation (i.e. with customers or suppliers) is most common in the manufacturing sector where 47% of firms cooperated with their customers and 46% cooperated with their suppliers. The data presented in Table 1.7 also shows that large firms are more likely to be involved in cooperation than small firms and that the gap between small and large firms in Ireland is much wider than in the EU generally. If, as is commonly accepted, innovation cooperation is linked with successful innovation, small Irish firms need to increase their involvement of external organisations in innovation projects. However some recent publications suggest that the link between innovation cooperation and success may not be as strong as previously assumed (Love & Roper, 2001; Brouwer & Kleinknecht, 1996; LaBahn et al. 1996).

Patent Applications

As the percentage of the labour force involved in R&D in Ireland approached the EU average one could expect that the output of R&D activity in Ireland would also approach the EU average. However, this was not the case. Patent applications are frequently used as a measure of the output of R&D activity. In 1998, 144 EPO (European Patent Office) patent applications were filed in Ireland. This compares with 4,850 applications in the UK. Even allowing for Ireland's smaller size these figures reflect a low level of R&D output, see Table 1.8.

	Ireland	**UK**	**EU - 15**
Number of patents in 1998	144	4,850	40,025
Per million inhabitants	27	78	88
Per million employed persons	82	174	222
Average annual growth rate 1989-1998	11.7	0.7	3.2

Source: Statistics on Science and Technology in Europe, EC, 2000

Table 1.8 Patent Applications –1998

In the UK, in 1998, there was a patent application for every 75 R&D personnel, in Ireland there was one for every 117 R&D personnel. The EU average for patent applications per million inhabitants was 88 while that in Ireland was 27. Based on these data, Ireland is classified as having a significantly lower technology density than other EU countries. To put this in perspective, in 1998 firms in East Anglia filed 424 patent applications and firms from the Oberbayern region of Germany filed 2,082 applications. Ireland's low level of patent applications is at odds with the number of Irish firms who claim to be involved in innovation. This could indicate that the innovation efforts of Irish firms are not as effective as those of firms from other EU countries. Alternatively, the use of patent applications as a measure of the innovation output for Irish firms could be questioned.

However, despite Ireland's poor performance in patenting, the number of patent applications in 1998 has risen considerably from the number of applications in 1988, when only 50 patents were filed in Ireland (Eurostat, 2000).

Across the EU it was also found that the proportion of innovating firms who applied for patents increased with firm size. In the manufacturing sector, 15% of small innovation enterprises applied for patents, compared with 51% of large enterprises. Not surprisingly it was also found that the number of patent applications was higher in the manufacturing sector (25%) than in the service sector (7%).

The low number of patent applications in Ireland could be related to the nature of R&D expenditure in this country. Recent figures from Forfás, the national science and technology agency, show that 84% of R&D expenditure in the electronics sector is on experimental development as opposed to basic or applied research which would be expected to result in patentable technologies. Irish electronics firms are more involved in developing new products than in developing new patentable technologies.

Turnover from New Products

Another measure of output for the innovation process is the amount of revenue that firms generate from new and improved products. This is shown in Table 1.9 for manufacturing enterprises in Ireland.

	Ireland	**UK**	**European Average**
% of Turnover due to New and Improved Products	32	23	32
Small (20 – 49)	21	14	15
Medium (50 – 249)	26	21	21
Large (250+)	43	25	38
NACE 29 - 33	69	44	45

Source: Statistics on Science and Technology in Europe, EC, 2000

Table 1.9 Turnover due to new and improved products - 1996

Again, the electronics sector is seen to be highly-innovative; in both Ireland and the UK it is the sector that generates the highest percentage of its turnover from new products. In fact, for the period in question,i.e 1996, the Irish electronics sector generated a higher percentage from new products than any other sector in any other EU country. Therefore, although Ireland performs poorly on one measure of the output of product innovation –i.e. patents, it performs well on another – i.e. revenue from new products. This could imply that much of the product innovation that takes place in Irish firms is incremental.

However, these results should not be taken at face value. Ireland's electronics industry is dominated by large multinationals that are known to take advantage of Ireland's low levels of corporate taxation by allocating a high percentage of their profits to their Irish manufacturing plants (i.e. they engage in transfer pricing). The result of this may be to artificially inflate the revenue generated in this sector. Transfer pricing has a significant impact on the level of export from the Irish electronics sector as will be seen in the next section.

Exports

The export intensity of innovating firms was also examined in CIS2. Table 1.10 shows the percentage of innovators among exporters and non-exporters.

	Ireland	**UK**	**European Average**
Non-exporters	46%	53%	40%
Exporters	78%	65%	57%
Total			
Of which:			
Low	70%	60%	52%
Medium	73%	64%	58%
High	83%	72%	61%

Source: Statistics on Science and Technology in Europe, EC, 2000
Note: Export intensity of measured as the ratio of export sales to total sales. Low: < 10%, Medium: 10% - 40%, High: >40%

Table 1.10 Export Intensity in Innovating Enterprises - 1996

These results show that in Ireland only 46% of firms who do not export are involved in innovation. At the other end of the scale, 83% of firms who export more than 40% of their total sales are involved in innovation.

In other words, firms that export are also more likely to be involved in innovation and the higher the export intensity, the higher the percentage of innovators.

What Table 1.10 does not show is the level of export from high-tech firms or its importance to the economy. Ireland is frequently described as a small open economy and the data presented in Table 1.11 support this description.

	Ireland	UK	EU-15
1995			
High-tech exports	11,961	38,096	87,735
As a % of total exports	35.0%	20.9%	15.3%
As a % of GDP	23.5%	4.4%	1.3%
High-tech imports	7,004	36,454	96,916
As a % of total imports	28.3%	17.8%	17.8%
High-tech trade balance	4,956	1,642	-9,181
2000			
High-tech exports	33,810	76,847	185,041
As a % of total exports	40.5%	24.9%	19.7%
As a % of GDP	32.7%	5.0%	2.2%
High-tech imports	20,685	81,446	233,892
As a % of total imports	37.4%	21.9%	22.6%
High-tech trade balance	13,125	-4,599	-48,851

Source: Statistics on Science and Technology in Europe, EC, 2000

Note 1: Figures given in million Euro.

Table 1.11 Trade in High-tech products in the EU – 1995 and 2000[1]

The importance of the high-tech sector to the Irish economy is apparent in the above data. In 2000 high-tech exports accounted for 40% of total exports and a third of GDP. Additionally, Ireland had the largest positive high-tech trade balance in the EU. While it is acknowledged that much of this is due to exports from large multinational firms whose profits were expatriated, this is still a critical sector of the Irish economy (Kearns & Ruane, 2001).

1.3 Emerging Questions

In Section 1.2 the electronics industry in Ireland was seen to be highly involved in innovation, spending a higher percentage of turnover on R&D than other manufacturing sectors. However, despite a rapid growth in R&D expenditure and in the number of R&D personnel in Ireland during the 1990s, both measures were still below the EU average by the end of the decade.

In terms of outputs from the innovation process Ireland is well below the EU average number of patent applications. However, when other measures of the output for innovation are examined, Ireland fares better. Irish electronics firms report that a very high percentage of their turnover is generated from new products. This could possibly suggest that Irish firms are good at incremental innovation or that they are not protecting the intellectual property that they develop.

The data presented above highlight the importance of the electronic manufacturing sector to the Irish economy and within this sector innovation (including R&D) has been shown to be a key activity in which firms invest heavily. Several issues arise from an examination of these data. Ireland is performing well in terms of inputs to the innovation process – many firms claim to be innovating, R&D investment, though still below the EU average, is increasing rapidly, and employment in R&D is increasing – yet when measured in terms of patent applications the output of Ireland's innovation is very poor. When the other two outputs of the innovation process are examined Ireland appears to be doing well but the degree to which this performance is due to the dominance of large multinationals within this sector must be examined.

Where data have been available to differentiate between small and large firms, small firms have been found to perform less well, both in terms of inputs to and outputs from the innovation process. Given the aim of this volume, i.e. to examine the management of R&D in small electronics firms, the above section has shown that, within the Irish context several, interesting issues arise.

1.3.1 Aims and Objectives

The above analysis of innovation within the electronics sector in Ireland raises several questions, for example:

- How well do small firms innovate?
- Should they increase their innovation expenditure?
- Would this improve their innovation outputs?
- How do small Irish firms actually manage NPD?
- How can small Irish firms improve their innovation performance?

Many other questions could be posed indicating that an *exploratory* approach to research in this area is appropriate. The aim of this research is to examine how small firms manage NPD but this examination is set against a background of extensive knowledge about how large firms manage NPD. Therefore, examining the differences between NPD in small and large firms is a useful approach to the research problem. The above questions suggest that it is necessary to examine more than *how* small firms manage NPD, examining how NPD performance can be improved is also important. Therefore the objective of this study can be achieved by answering the following research question.

How do small Irish electronics firms manage the development of new products, and are the success factors that apply to small Irish firms different from those that apply to large Irish firms?

1.4 Definitions

The term NPD when used throughout this volume refers to the process used by companies to develop new products. This can be defined as 'the range of activities and decisions from the time when an idea is generated (from whatever source) until the product is commercialised (i.e. launched onto the market)' (Caffyn, 1998, p.64). Within the literature published on the management of NPD upon which this study is based the terms 'product development', 'product innovation' and 'innovation' are widely used. Hart (1996), in the introduction to her reader on New Product Development comments that 'while there are real nuances of difference amongst these terms ... the basic concerns that fall within the domain of each are remarkably similar' (Hart, 1996, p. xi). She cites the example of a well-known article by Maidique and Zirger (1984) that examined the electronics industry and used the terms 'new product development' and 'innovation' interchangeably. Several other key authors in the field, for example Cooper and Crawfold, also refer to the process of bringing new products to the market as either 'new product development' or 'innovation'. In this case, innovation, defined by Tushman & Nadler (1987) as 'the creation of any product, service, or process which is new to a business unit' (p. 75), will be interpreted to mean product innovation, i.e. the creation of a product that is new to a business unit. This definition is clearly consistent with the definition of NPD given above.

This volume, as suggested by its title, is about the management of new product development (NPD). However, in reviewing literature that informs this study there will be times when the terms 'innovation' and 'product innovation' will be used to accurately reflect the content of a particular journal.

1.5 Overview of Volume

This section presented a brief overview of the structure of this volume.

Chapter 2 contains a review of the literature on NPD, concentrating on publications that examine determinants of new product success and highlighting references to small firms. **Chapter 3** reviews the publications that have resulted from an international study of the determinants for new product success - the INTERPROD project. This project is identified in Chapter 2 as being useful in answering the research question guiding this research.

The research model that will guide the project is presented in **Chapter 4**; the variables that populate the model are described, and the possible impact of firm size on these variables is noted. **Chapter 5** provides an overview of the research plan–i.e. a multi-method approach. The objectives of the research strategy, how it will be implemented, and its strengths and weaknesses will be described. The research question will be extended in order to make it operational.

A 'first cut' of the analysis of the Irish INTERPROD data will be presented in **Chapter 6**. This will involve comparing the Irish results with those presented in 5-i.e. previously published INTERPROD papers that were considered to be relevant to the Irish situation. This will serve to place NPD management in Ireland within an international context.

A detailed analysis of the Irish and UK INTERPROD data is presented in **Chapter 7;** these results are then discussed in **Chapter 8.** In these two chapters particular attention is paid to differences between NPD management in small and large, Irish and UK firms. In order to explore some of the findings of Chapters 7 and 8 in more detail three case studies were conducted. **Chapter 9** presented these three cases and they are discussed in **Chapter 10.**

The research model that guides this study is initially presented in Chapter 4. As the study progresses the model is updated in Chapters 6 and 8. The final version of the model is presented in Chapter 10 along with twelve hypotheses derived from the key research findings. Finally, **Chapter 11** contains the conclusion of the study.

1.6 Conclusion

This chapter presented the introduction to this volume. The first section introduced the research topic, the management of NPD in small Irish electronics firms. The background of the firms to be studied in term of their innovation performance was given in the second section. Based on several issues that emerged from this background information, the research question was formulated. The final section contained an overview of the structure of this volume.

CHAPTER TWO

LITERATURE REVIEW

2.1 Introduction

Studies of new product development have taken several different approaches. Some have compared successes and failures; others have concentrated on the impact of particular variables on NPD success; yet others have examined the NPD process and suggested ways in which it can be improved. The purpose of this chapter is to present a review of the literature on new product development as it is related to the research question posed in Chapter 1:

How do small Irish electronics firms manage the development of new products, and are the success factors that apply to small Irish firms different from those that apply to large Irish firms?

There are two parts to this question, firstly in managing the development of new products, what are the *practices* in small firms and secondly, are the *success factors* in small firms different from those in large firms. Both parts of this question are represented in the proposed model shown in Figure 2.1.

The aim of this research, and the objective of the proposed research model, is to find out how small Irish electronics firms manage new product development and determine what success factors apply. Literature on the management of NPD in *small* firms is scarce while there is a considerable body of knowledge about the management of NPD in *large* firms. Therefore, the approach taken to reviewing the literature in this chapter is to examine the determinants of NPD success in the general literature (which deals primarily with large firms) and to explore how it applies to small firms. This approach ensures that a comprehensive review of the determinants for NPD success is possible and also makes it possible to easily identify gaps in the literature relating to small firms.

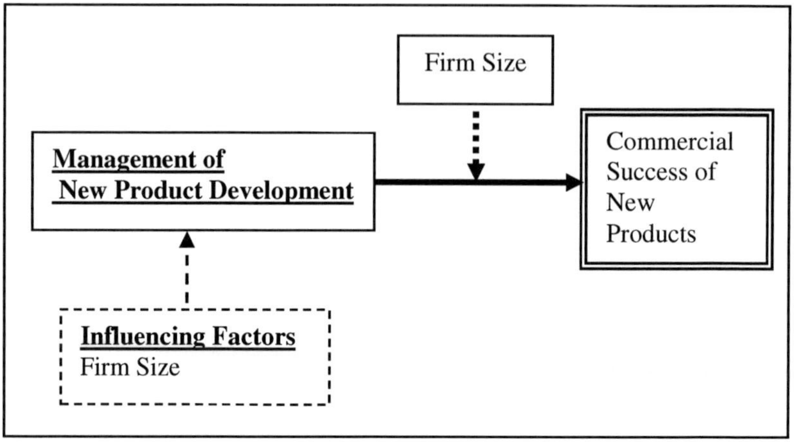

Figure 2.1 Research Question

The research model will be developed throughout this volume by identifying the NPD management practices that have been linked with new product success in large firms and exploring how they are affected by firm size. To this end, the first version of the research model, Figure 2.1 suggests that firm size has an impact on the way in which new product development is managed within a firm. It also suggests that firm size has an impact on the relationship between different NPD management practices and the commercial success of new products.

This chapter is laid out as follows:

- Section 2.2 presents some models of NPD success; these are used to identify the different management practices (variables) that will be examined during the course of this study.
- Sections 2.3, 2.4 and 2.5 examine these variables in more detail reporting the findings of previous studies and highlighting any references to small firms.
- Section 2.6 contains a summary of the literature review presented in Sections 2.3, 2.4 and 2.5.
- Section 2.7 presents the conclusion of this chapter.

2.2 Models of NPD Success

There are many models of the NPD/product innovation process (Rothwell & Zegveld, 1985; Tushman & Nadler, 1986; Chiesa et al. 1996; Tang, 1998; Tuominen et al. 1999) ;however, the objective of this study is not to develop a model of product innovation but rather to determine the factors that impact on new product success in small firms. Therefore the models presented in this section are those that link management practices with new product outcomes.

Cooper & Kleinschmidt (1987) developed an early model of the factors that influence new product outcomes. It was based on a set of nine hypotheses; a simplified version of the Cooper & Kleinschmidt model is shown in Figure 2.2. Cooper & Kleinschmidt found strongest support for H1, H5 and H6. In other words, there is a strong link between product advantage and the commercial success of a new product. Product advantage is achieved by having a clear definition of customer requirements, and product definition, and also by undertaking proficiently a set of pre-development activities including technical assessment, market research and business analysis. An additional factor, i.e. market competitiveness, was tested but not found to be related to new product success.

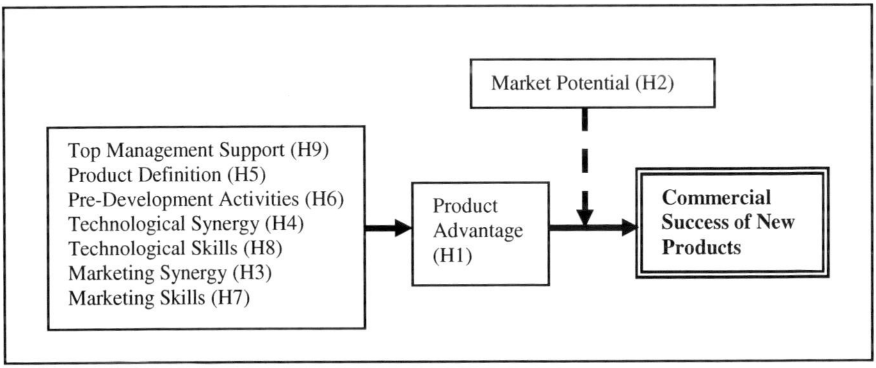

Figure 2.2 Cooper & Kleinschmidt's Model of New Product Outcomes

Brown & Eisenhardt (1995) developed a model of the factors that affect the success of product development; a simplified version of this model is shown in Figure 2.3.

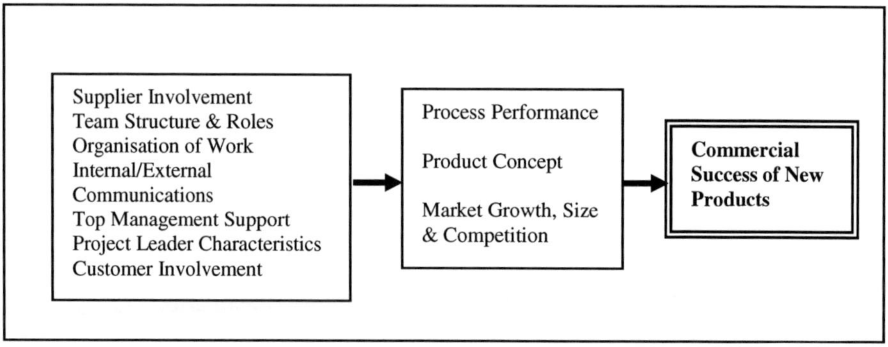

Figure 2.3 Brown & Eisenhardt's model of Factors Affecting New Product Success

This model results from the synthesis of three different streams of research:

- **Rational Plan**: Successful product development is achieved through '*careful planning of a superior product for an attractive market, the execution of that plan by a competent and well-coordinated cross-functional team that operates with the blessing of senior management*' (p.348).
- **Communications Web**: A successful product development process is dependant on good communications between project team members and also with key outsiders.
- **Disciplined Problem Solving**: Successful product development is a '*balancing act between relatively autonomous problem solving by the project team and the discipline of a heavyweight leader, strong top management support, and an overarching product vision*' (p.359).

There is a similarity between the models proposed by Cooper & Kleinschmidt and Brown & Eisenhardt. This is not surprising since Cooper was a significant contributor to one of the streams of research, the *rational plan*, upon which this model is based. The influence of the other two streams of research can be seen in the inclusion of communications and team role variables from the *communications web* stream of research and the supplier involvement and organisation of work from the *disciplined problem solving* stream of research.

Four publications, (Song, Souder & Dyer, 1997; Souder & Song, 1998a; Souder & Jenssen, 1999; Lee, Lee & Souder, 2000) all based on the same data set – INTERPROD (an international study into the factors that influence new product success) – also present models of the factors influencing new product success. These are presented in Figure 2.4, modified to reflect the findings of each set of authors. Again a range of management practices is found to have an impact on the commercial success of new products.

All of the models presented so far show that new product success is related to the management of the NPD process. Several models, particularly those developed by Cooper (1987) and Brown & Eisenhardt (1995) attempt to explain the interrelationships between management variables and NPD process characteristics. For example, both Cooper and Brown & Eisenhardt relate top management support to effective new product strategy, which in turn is related to new product success. The approach taken in the models presented in Figure 2.4, and by this study, is simpler examining only the direct links between management practices and new product success, rather than interrelationships between different management practices.

Several of the models also suggest that market related (external) factors such as market potential, growth, size, competitiveness and familiarity have an impact on the success of new products. The objective of this study is to identify the relationship between (internal) NPD management practices and success in small Irish electronics firms. It is acknowledged that market related factors are important, but they are not the focus of this research. Additionally, as the firms involved in this study all come from the same industrial sector, electronics, they are all operating within broadly similar markets. This should limit the variability of market related factors within this study.

National setting is an external variable included in two of the INTERPROD models (Souder & Jenssen, 1999; Lee, Lee & Souder, 2000). The focus of this study is on Irish firms but it may be useful to examine the extent to which the differences found between small and large firms in the Irish data are also seen in other countries.

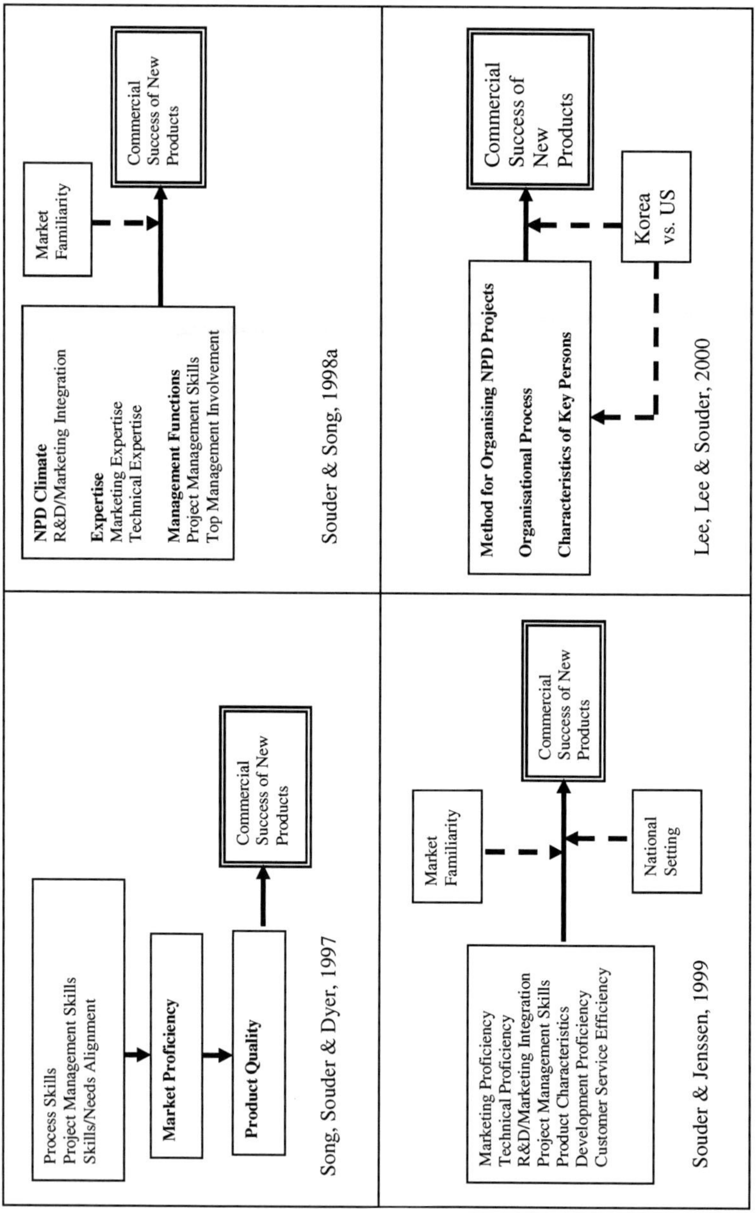

Figure 2.4 Summary of INTERPROD Models

The model derived from the research question, see Figure 2.1, can be seen as a simplified overview of the models presented above. In common with the other models the general proposition is that new product success is dependant on the way in which NPD is managed. However the model in Figure 2.1 includes firm size as a key variable, suggesting that NPD management and new product success are both dependant on firm size. Having presented a general overview of the model to be developed the following section identifies the NPD management variables that will populate this model.

2.2.1 Determinants of NPD Success

The models presented above link NPD management practices with new product success. The purpose of this section is to summarise the variables considered in these models into a single list that avoids duplication but includes all relevant factors, particularly those that have been found to be influenced by firm size.

Montoya-Weiss and Calantone (1994) provide a comprehensive analysis of research into NPD and conclude that it lacked focus and direction. A wide variety of methodologies and study types have been used making it difficult to compare the results of different studies (Holger, 2002). However, in their meta-analysis they define eighteen variables in four categories that have been identified as determinants of new product performance, Table 2.1

As explained above market environment factors are not of key importance to this study, also many management decisions relating to the market environment for new products can be considered to be strategic. Therefore, strategic and market environment factors will be considered together so that the variables to be examined in this chapter will be organised under three headings:

- Strategic and Market Factors
- Development Process Factors
- Organisational Factors

Strategic Factors	*Development Process Factors*
Product Advantage Technological synergy Marketing synergy Company resources Strategy	Protocol Proficiency of Predevelopment Activities Proficiency of Market-related Activities Proficiency of Technological Activities Top management support/skill Speed to market Costs Financial/Business Analysis
Market Environment Factors	***Organisational Factors***
Market potential Market competitiveness Environment	Internal/External relations Organisational factors

Adapted from Montoya-Weiss and Calantone, 1994

Table 2.1 Determinants of NPD success

These three headings have been used to structure the variables from each of the models presented in Section 2.2, see Table 2.2.

The final section of Table 2.2 lists the variables to be examined in this literature review. The variables have been selected to include those examined in the previous models but to avoid duplication. In addition, two other factors are included in the final section of Table 2.2; these are Cycle-Time Practices and Resources for NPD.

The inclusion of cycle-time practices as a factor is because of the amount of literature that suggests that the ability to develop new products rapidly is linked with success (Ali et al. 1995; Eisenhardt & Tabrizi, 1995; Zirger & Hartly, 1996; Griffin, 1997; Callahan & Moretton, 2001). Resources for NPD is included as a variable because it has been identified as an issue for small firms (Roper et al. 1996; Hoffman et al. 1998; Love & Roper, 2001).

The focus of this thesis is on the management of new product development in *small* firms, however literature in this area is scarce (March-Chordà et al. 2002, Berry & Taggart, 1998, Acs & Audretsch, 1988). Therefore, the next three sections will report on the general literature in each category defined in Table 2.2 and highlight any references to small firms. Section 2.2 at the end of this chapter will summarise the publications and findings that relate to small firms.

The next three sections in this chapter address each of the three groups of factors influencing NPD success:

Section 2.3 Strategic and Market Factors
Section 2.4 Development Process Factors
Section 2.5 Organisational Factors

Strategic and Market Factors	**Development Process Factors**	**Organisational Factors**
Cooper (1987)		
Product Advantage Market Potential Product Definition Technological Synergy Marketing Synergy	Pre-Development Activities Top Management Support Technological Skill Marketing Skill	
Brown & Eisenhardt (1995)		
Market Growth Size and Competition Product Concept	Process Performance Team Structure and Roles Top Management Support Project Leader Characteristics	Supplier Involvement Customer Involvement Organisation of Work Internal/external Interaction
INTERPROD Papers Song, Souder & Dyer (1997)		
Skills/Needs Alignment Product Quality	Process Skills Project Management Skills Marketing Proficiency	
Souder & Song (1998)		
Market Familiarity	Marketing Expertise Technical Expertise Project Management Skills Top Management Involvement	R&D/Marketing Integration

Souder & Jenssen (1999)		
Market Familiarity National Setting Product Characteristics	Marketing Proficiency Technical Proficiency Project Management Skills Development Proficiency Customer Service Efficiency	R&D/Marketing Integration
Lee, Lee & Souder (2000)		
	Characteristics of Key Persons	Methods of Organising Organisational Processes

Variables to be included in this study:

Strategic and Market Factors	**Development Process Factors**	**Organisational Factors**
New Product Characteristics Market Characteristics Marketing and Technological Synergy Resources for NPD	Top Management Support Marketing and Technological Skills Development Process Proficiency Project Management Skills Cycle-time Practices	Involvement of External Organisations R&D/Marketing Integration Methods of Organising NPD

Table 2.2 Summary of Model Variables

2.3 Strategic and Market Factors

The strategic management of NPD is a key determinant of new product success as it affects all other factors. Many of the strategic issues that concern large firms also apply to small firms (Dodgson and Rothwell, 1991). Learning, flexibility and speed of response are identified as areas where small high-technology firms can gain a competitive advantage. Adler et al. (1989) suggest that most companies manage NPD in a tactical rather than a strategic manner, top managers are advised to take a more strategic interest in NPD. They identify five elements of technology management that should be managed more strategically; these are - the assignment of management responsibilities, the role of functional departments, the nature of planning, protecting competitive advantage and the link between technology and new product management. Many of these issues have since been developed by other authors (Clark & Wheelwright, 1992; McGrath, 1995; Griffin & Hauser, 1996; Swink, 2000).

Cooper & Kleinschmidt (1995) proposed a framework for a company's overall NPD performance consisting of five elements. They identified the *main performance drivers* that separated high NPD performers from other firms, in order of importance these were:

1. A high-quality NPD *process* that was executed thoroughly and in a quality fashion.
2. A NPD *strategy* should be clearly stated and well communicated.
3. Senior management *commitment.*
4. A firm should have an entrepreneurial *culture.*
5. A suitable *organisation* for NPD projects.

Cooper & Kleinschmidt show that strategy is a determinant of new product success but say nothing about what type of strategy is appropriate in different situations. Pearson (1991) used technical and market uncertainty to describe a matrix defining types of development activities each with different strategic implications. Clark (1983) took a similar view examining the impact of product innovation on a firm based on the degree to which it requires new technical or marketing skills. These models are combined in Figure 2.5.

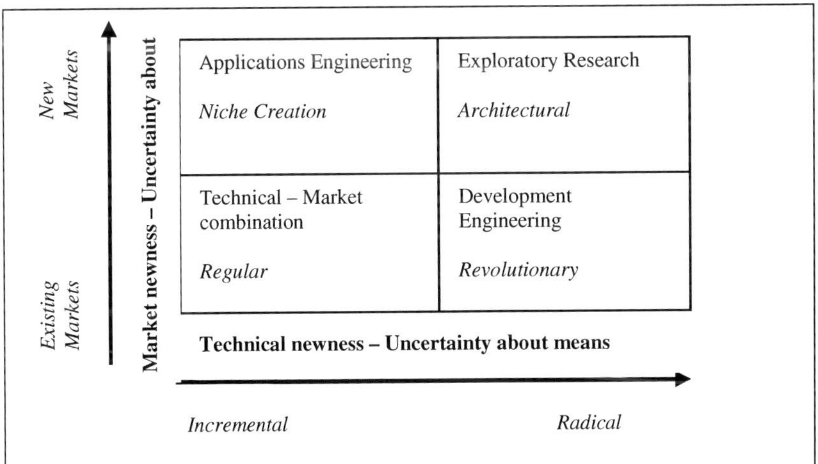

Adapted from Pearson (1991) and Clark (1983)

Figure 2.5 Technical and Market Uncertainty

The uncertainty map shown in Figure 2.5 suggests that new product strategy is dependent on both the nature of the technology to be employed and the nature of the market to be addressed. These two issues and their impact on new product strategy and success are addressed in the following four sections:

Section 2.3.1	New Product Characteristics
Section 2.3.1	Market Characteristics
Section 2.3.3	Marketing and Technological Synergy
Section 2.3.4	Resources for NPD

2.3.1 New Product Characteristics

Cooper (1994), points to product superiority or advantage, that is the advantage of the product in the eyes of the customer, as the most important key to new product success, above any technical or market environmental issues. This is supported by others. A comparative study of the food processing and medical instrument industries (Karakaya and Kobu, 1994) identified poor product performance as a common cause of failure. Cooper (1990) reporting on the project NewProd, concluded that superior products are more likely to succeed. However, the Yap and Souder (1994) study found that small firms, under conditions of high market or technical uncertainty, were more likely to succeed by producing compatible rather than superior products.

Another new product characteristic that has been found to have a significant impact on new product success is the extent to which radical, as opposed to incremental, new technology has been employed. Walsh et. al. (2002) surveyed 72 micro-electro-mechanical-systems manufacturing firms to determine the extent to which small firms favour the commercialisation of *disruptive* as opposed to *evolutionary* technologies. Disruptive technologies are described as those that are new both to the firm and to the firm's customer base – these would occupy the top right quadrant in Figure 2.5, *architectural* innovation. Evolutionary technologies are defined as those that provide incremental improvements building on existing technologies – the bottom left quadrant in Figure 2.5, *regular* innovation. Walsh et. al. equated firm size with firm newness and defined firms with fewer that 100 employees as being new firms. They found that new firms did not show a preference for disruptive technologies but that they did have the advantage of being able to get products to market in significantly less time than established firms. They concluded by suggesting that the behaviour of new (in reality small) firms is not fully understood and requires further research.

An earlier study by Santarelli & Sterlacchini (1990) that examined Italian manufacturing firms also found that small firms introduced mainly incremental product innovations. In addition they noted that R&D intensity was higher in small firms, a finding supported by other authors (Acs and Audretsch 1988; 1990; Van Dijk et. al., 1997; Love and Roper, 2001). A preference for incremental product innovation is found not only in small firms, large, established firms also have difficulties in successfully commercialising disruptive technologies (Bower & Christensen, 1995).

A strategy recommended for small technology based firms is the adoption of one key growth-sustaining technology and avoidance of high levels of diversification (Meyer and Roberts, 1986). Successful initial core technologies are those that are challenging to implement but difficult enough to deter competitors. Small, high-technology firms are advised to avoid differentiation but to adopt cost-leadership or focus strategies (Yap and Souder, 1994). This advice is based on empirical research that shows that customers are unlikely to buy superior products with unique features from new companies. Hoffman et al. (1998) also suggest that small firms lack the credibility and track record to be successful with truly radical products.

2.3.2 Market Characteristics

Several aspects of the market into which new products are launched have been considered as success factors. An early study found that environmental variables, including market potential, did *not* play a crucial role in NPD success (Cooper, 1979). But later studies show that market potential, including market size and growth, customer need for a product and the importance of a product to the customer, has an impact on NPD success, particularly where companies are aiming their product at a window of opportunity (Cooper and Kleinschmidt, 1987).

Acs & Audretsch (1988) concluded that small firms had an advantage in newer markets, or industries that were in the early stages of their life cycle. On the other hand, large firms had an innovative advantage in markets that were capital intensive, concentrated and advertising-intensive. This supports the Abernathy & Utterback (1978) model of industrial innovation. During the early or fluid stage of an industry's development the flexibility of small firms gives them an advantage that they lose as an industry matures towards its specific stage when large firm's economies of scale are more appropriate. In 1988, Acs and Audretsch found that small firms were more innovative within the electronics sector. However, as this industry has matured the position of small firms is less clear.

It is a common assumption, based on Porter (1980), that increased market competitiveness should limit the success of new products. However several authors have not found this to be true (Booz-Allen & Hamilton 1982; Cooper and Kleinschmidt 1987; Page 1993). In the case of small firms there is no consensus on the impact of market competitiveness on success, some studies have found it to correlate negatively with success while others have found that it has no impact at all (Yap & Souder 1994; Hoffman et al. 1998).

Meldrum & Millman (1991) identify ten specific risks in marketing of high technology products including; inadequate technology, specification drift, technology 'leapfrogging', credibility, time/cost overruns and standards. While all high-technology firms have to manage these risks some pose particular problems for small companies. New and rapidly changing technologies are frequently accompanied by fast-growing or collapsing markets (McGrath, 1995; Shanklin & Ryans, 1984), smaller firms will not always have the resources to keep up. Yap & Souder (1994) found that small, high-technology companies are subject to different constraints and opportunities than larger firms. Their data linked market maturity to product success. They explained this unusual result by pointing out that many of the small companies they studied were in small mature market niches that were not of interest to larger firms.

2.3.3 Marketing and Technological Synergy

Technological and marketing synergy were linked with new product success in an early study of the U.S. electronics industry (Maidique & Zirger, 1984). Maidique & Zirger measured the importance to firms of different determinants of NPD success and found that environmental factors including; government regulations, market characteristics and the economy, did not rate very highly. They concluded that seven factors had a positive impact on the success of NPD. One of these was that; 'the markets and technologies of the new product benefit significantly from the existing strengths of the developing business'. Companies are advised to 'attack from a position of strength' (Cooper, 1994). This advice has also been shown to be true for small firms (Yap & Souder, 1994).

More recent studies also support the importance of marketing and technological synergy to new product success (Song & Parry, 1999; Balbontin et al. 1999). In summary, there is general acceptance within the literature that NPD projects that are aligned with a firm's existing marketing and technological skills have a greater chance of success than those that are not. But there are very few publications that examine the impact of synergy on new product success in small firms.

2.3.4 Resources for NPD

The availability and quality of resources as a determinant of NPD success has not been studied to any great extent. It is often assumed that NPD programmes are sufficiently resourced; this may well be the case in large firms. A study of the norms in product development (Page, 1993) found that resources were mentioned by 39.2% of 189 companies surveyed as being obstacles to successful NPD. In small firms, a lack of resources, both financial and human, has also been cited as one of the most important barriers or constraints to product innovation (Fitzgerald & Breathnach, 1994; Roper et al., 1996; Hoffman et al., 1998).

Almost a third of Irish manufacturing firms surveyed by Roper (1998) indicated that 'their innovation activities had been hindered by skill shortages' (p.371), this compared with only 22% of German firms. Roper also found that the number of manufacturing plants introducing new or improved products in Ireland (61%) was lower than in Germany (71%). Roper's publications are based on the Product Development Survey that was conducted in Germany, Ireland and the UK between 1994 and 1995. His sample included 533 Irish firms, 1374 German firms and 1722 UK firms. In a paper that concentrated on small firms (Roper, 1997) he showed that the most important objectives for small firms (between 20 and 100 employees) in innovating were raising product quality and extending the product range. He also reported that growth in turnover was much faster in innovating firms than in non-innovators. Roper found similar characteristics in the UK and Irish firms and frequently discussed both groups of firms together. He noted that R&D in small Irish and UK firms was informal and flexible, and that these firms undertook product innovation often in response to market opportunities. By contrast the

German firms were found to have a more systematic and planned approach to developing new products.

A study of product development in 74 Japanese electrical machinery firms (Wakasugi and Koyata 1997) found that while the number of new products developed and the number of patent applications increased with firm size there was no evidence of economies of scale in relation to NPD. In other words, small firms can be as efficient in their product development activities as large firms.

In summary, strategic factors - product advantage, technical and market synergy, resources and company strategy - are determinants of new product success for both small and large firms. However, several authors have identified differences between how strategy should be planned and implemented in small firms in order to increase the probability of new product success.

2.4 Development Process Factors

The process used to develop new products has been examined in detail by many authors (for example Cooper & Kleinschmidt, 1986, 1987; Song et al. 1998; Song & Montoya-Weiss, 1998; Veryzer, 1998). Some studies have concentrated on the overall structure of the NPD process, the benefits of a stage-gate strategy over a phase-review system or indeed, whether a formal process is required at all. Other studies have focused on the individual steps within the process that are most closely linked with success and also on ways to speed up the process. Finally, several authors have taken a contingent approach claiming that there is no 'best' process for developing new products but rather that the process should be selected to suit the context within which the firm operates (Bessant & Francis, 1997; Loch, 2000).

Roper (1997) found that in small Irish and UK manufacturing firms R&D tended to be informal and often in response to market opportunities. This finding is supported by Hoffman et al. (1998) who reported that R&D activities 'tend to be more ad-hoc or project driven in smaller SMEs'. Santarelli & Sterlacchini (1990) showed that 'informal R&D is an important part of the total R&D preformed by small and medium sized firms' and they suggested that 'systematic R&D undertaken by large firms within structured laboratories is more effective than occasional R&D carried out by small firms'.

One factor that has been found to have an impact of the type of NPD process employed is the level of innovativeness, or newness of the product being developed. Rochford & Rudelius (1997) reported that new-to-the-world products are normally developed using a more complete NPD process than product modifications. They also found that there was a greater emphasis on preliminary market analysis activities for really new products, this result is supported by Song & Montoya-Weiss (1998). However, not all researchers support the use of formal, stage-gate processes for very innovative products. Veryzer (1998) examined the processes used by firms developing radically new products. Using data from eight case studies, he concluded that 'Stage-gate-type methods, which are utilised for continuous products being developed by these firms, have not proven to be conducive for the development of discontinuous products' (p.318). He recommends the early development of prototypes and flexibility in the application of a formal NPD process when developing discontinuous new products.
The above examples show that despite conflicting evidence about the *type* of development process most closely linked with new product success, there is agreement that the development process employed has a significant impact on the outcome of a NPD project.

The review of literature relating to the development process factors that determine new product success will be divided into five sections:

Section 2.4.1 Top Management Support
Section 2.4.2 Marketing and Technological Skills
Section 2.4.3 Development Process Proficiency
Section 2.4.4 Project Management Skills
Section 2.4.5 Cycle-Time Practices

2.4.1 Top Management Support

Top management support has been identified as a NPD success factor by several empirical studies (Cooper & Kleinschmidt, 1995; Swink, 2000) and by two comprehensive reviews of NPD literature (Montoya-Weiss & Calantone, 1994; Brown & Eisenhardt, 1995). In the US electronics industry senior management support has been linked with NPD success (Maidique and Zirger, 1984). Owner manger skill and support has also been found to be important in small firms (Hoffman et al. 1998) though the focus when dealing with small firms is more on the skill and competence rather than the support of top management. This is justifiable as it would be very unusual for a small firm to take on a product development project without the support of top management (Maidique, 1980). Top management support has been found to vary in different cultures with Kleinschmidt (1994) reporting that top management in Europe is more involved in NPD that in North America.

Swink (2000) examined the impact of top management support (TMS) on NPD goal achievement in 137 manufacturing firms in the US. Swink proposed that TMS can have several positive effects on NPD, these include:

1. It can act as a force to pull together different functional groups.
2. TMS generates more interest, enthusiasm and greater ownership in projects.
3. Top managers can quickly acquire unanticipated project resources.
4. Designers pay more attention to details and worry less about resources required for analyses and testing when there is TMS.
5. Top managers provide a clear vision of the product concept and goals.

However, Swink does warn that there is a stage where top management support can turn into interference, and this can be especially detrimental in high technology, innovative projects (Takeuchi & Nonaka, 1984). Swink found that projects with high levels of TMS had better time-based performance, design quality, and financial performance. He also found that the impact of TMS was greater in low technology than in high technology projects and that TMS was not associated with financial success in very high technology projects. Swink suggests that the reason for this is that in high technology projects top managers are more likely not to understand many of the issues involved and so to make incorrect decisions.

2.4.2 Marketing and Technological Skills

A link between both technological skill and marketing skill and new product success was established in early studies (Cooper, 1987; Cooper & Kleinschmidt, 1987; Zirger & Maidique, 1990). Since this is such an intuitively logical link it has not been re-examined in many more recent studies. An exception to this was a study by Wren et al. (2000) that examined determinants of success in six countries and found that technical skills were more important in countries that have a disposition toward exporting and marketing skills were not linked with success in any of the countries studied. This is a surprising result and contradicts some of the earlier findings of Cooper who would have suggested that marketing skills are more critical to new product success than technical skills.

2.4.3 Development Process Proficiency

There is wide agreement that a formal NPD process is necessary to effectively manage the development of new products (Rochford & Rudelius, 1997; Bessant & Francis, 1997; Cooper & Kleinschmidt, 1986). Early studies of the NPD process concentrated on the stages involved and the link between proficiency at different stages and new product success (Cooper, 1979; Cooper & Kleinschmidt, 1986, 1987). These studies showed that successful new products tended to be developed using processes that featured more NPD stages than unsuccessful products (Cooper & Kleinschmidt, 1986). More recent research has supported this finding; Rochford & Rudelius (1997) found that 'more successful new products have more complete NPD processes than less successful new products'. Rochford & Rudelius also reported that new-to-the-world products are normally developed using more complete NPD processes than product modifications. This contradicts the findings of Veryzer (1998) discussed earlier in Section 2.4.

The question of which NPD process stages are the most important was also addressed by the early Cooper & Kleinschmidt (1986, 1987) studies. Predevelopment activities and technological activities were both found to be linked with new product success. Song & Montoya-Weiss (1998) examined the impact of proficiency of six sets of NPD activities on new product success for 'really' new products and incremental products. They had expected to find that post development activities (market opportunity analysis and commercialisation) would be important for incremental projects and technical development and testing would be important for really new products but this was not the case. Instead they found that commercialisation was the most important activity for really new products while market analysis was most important for incremental products. These results show that the level of innovation of a new product has an impact on the type of process that should be applied in terms of which activities are the most important.

Loch (2000) also questioned the appropriateness of using the same NPD process in all situations. He studied a European technology manufacturer who had a 'standard' NPD process; very different processes were actually applied in different situations. Loch proposes that the choice of the 'best' process is dependent on the strategic innovation needs of the business unit developing a new product. The importance of how a NPD process should be implemented to suit a particular firm is stressed by Bessant and Francis (1997) who point out that, 'what works for one firm cannot simply be copied by another with the same results'. They emphasise the fact that putting in place a NPD process is an organisational development activity that should be managed in an inclusive manner to ensure ownership of new process.

2.4.4 Project Management Skills

Early studies have indicated that a 'strong' project manager is linked with success (Clark & Fujimoto, 1991; Wheelwright & Clark, 1992). In this context a strong project manager, also called a heavyweight project manager, is defined in terms of his/her position within the organisation i.e. strong manager is one in a senior management position. Another stream of research proposes that a project manager's behaviour and competence rather than power is an important factor in new product success (Norrgren & Schaller, 1999).

Several studies have examined project manager's technical skills, marketing skills, management skills, participative style and authority from top management (Yap & Souder, 1994; Yap et al., 1998; Souder & Song, 1998; Souder & Jenssen, 1999; Lee et al. 2000). For example, Table 2.3 shows the results of Yap et al. (1998) who studied 24 projects in Singapore based electronics firms and concluded that 'the skill base of the project manager is critical for project success'.

Item	**Mean Scores (1-5 Scale)**	
	Success (n = 14)	**Failure (n = 10)**
Technical Skills	4.4	3.2
Marketing Skills	3.8	2.4
Management Skills	4.1	2.9
Participative Style	4.4	3.3
Authority from Top Management	4.4	3.6

Table 2.3 Impact of Project Management Skills on Project Outcomes

The other studies cited above reached similar conclusions, however the importance of Project Management Skills was found to be different in different countries. Souder & Song (1998) concluded that technical skills were more important for project managers in Japan than in the US, while Lee et al. (2000) found that a project manager with a participative style was more closely linked with success in Korea than in the US. Souder & Jensen (1999) found that project manager competency was more important under conditions of high market uncertainty. Finally, Yap & Souder (1994) found that in small electronics firms having a project manager with a participative style was *not* linked with new product success.

In summary, several studies have established a link between project manager's skills and new product success but the nature of this link has been found to be dependent on national setting, market uncertainty and firm size.

2.4.5 Cycle-time Practices

Several studies have explored different strategies for reducing the time it takes to develop new products. Eisenhardt & Tabrizi (1995) recommend an experiential strategy, based on frequent design iterations, flexibility and rapid learning, for reducing the NPD time of *uncertain* projects. But they found that a compression strategy, based on carefully managing the steps involved in NPD and where possible shortening them or conducting several steps in parallel, was effective for reducing the NPD time of *certain* projects

Zirger & Hartley (1996) advise that 'Faster development can be achieved by increasing the number of functions represented on the team, reducing the number of projects to which team members are assigned and placing a high priority on time as a goal'. While, Griffin (1997) suggests that – 'the length of the product development cycle is always related to the product complexity and newness of projects' and 'both process related factors (cross-functional teams and a formal NPD process) reduce cycle time'. Finally, Datar et al. (1997) found that a *concentrated structure*, i.e. one where all product developers are in the one location, facilitates faster prototyping while a *distributed structure*, i.e. one where the NPD effort is dispersed among several manufacturing locations, facilitates faster time to volume production.

The previous four papers (Eisenhardt & Tabrizi, 1995; Zirger & Hartley, 1996; Griffin, 1997; Datar et al., 1997) all examined the development of hardware products. Callahan & Moretton (2001) looked at planning software development and found that additional planning did to reduce development time, longer projects required more planning but there was no evidence that increasing planning as a proportion of total development time reduced the overall duration of a project.

Dröge et al. (2000) suggest that the use of CAD is the practice most strongly correlated with the ability to minimise NPD times. Dröge et al. also examined the issue of firm size and found that supplier development, JIT purchasing and concurrent engineering all increased with firm size. Ali et al. (1995) studied the impact of various factors on cycle time in *small* firms. They suggest reducing cycle time by developing incremental rather than very innovative products or products contain state-of-the-art technical know-how and by concentrating on products that offer genuine advantages to the customer.

There are two interesting things to note about the papers reviewed above. Firstly, there is strong evidence that the type of product being developed, in terms of novelty, certainty and complexity, must be considered when selecting a strategy for reducing time-to-market (Griffin, 1997; Callahan & Moretton, 2001). Secondly, the papers reviewed in this section have not stressed the link between time-to-market and new product success but have relied on earlier studies (Vesey, 1991) to make this connection. Cooper (1994) warns that although being first to market is important being the best on the market is more important.

2.5 Organisational Factors

The management of internal and external relationships in new product development is one area where there is a marked difference between the treatment in literature of small and large firms. Most of what is published about large firms deals with internal relationships. It has been argued that functional integration between departments will achieve better results both in the characteristics of the products developed and the time taken to develop them (Shrivastava and Souder, 1987; Wheelwright and Clark, 1992; Towner, 1994). In contrast, literature dealing with small firms focuses on external relationships with issues such as industrial services, subcontracting relationships, licensing, networking, collaborative R&D (Rothwell and Dodgson, 1991; Hoffman et al., 1998; MacPherson, 1997). This difference in emphasis makes some sense - large firms tend to be able to resource all their technical requirements internally and thus the relationships the must manage are internal. Small firms however often need to look outside their organisational boundaries for services and therefore must develop skills in managing external relationships.

Organisational factors including the existence and characteristics of project teams, the level of top management involvement, the nature of leadership of NPD projects, have been examined extensively in large firms. Cross-functional or multidisciplinary teams with strong leadership and a high level of ownership and accountability are generally recommended as good organisational design for new product development projects (Wheelwright and Clark, 1992; Page, 1993; Cooper, 1994).

This section will examine three organisational factors that have been identified within the literature as having an impact on the success of new products. These are:

Section 2.5.1 Involvement of External Organisations
Section 2.5.2 R&D/Marketing Integration
Section 2.5.3 Methods of Organising NPD

2.5.1 Involvement of External Organisations

It is generally perceived that it is good for firms (especially small firms) to involve external enterprises in their NPD efforts. NPD cooperation is encouraged within EU policy - 'Collaboration with different enterprises results in lowering the costs and risks of innovation as well as sharing scientific and technical knowledge' (Eurostat, 2000). However, not all studies support a strong link between networking and the level of product innovation. Freel (2003) explored the relationship between networking and innovativeness and reported that positive links were only found for firms cooperating with customers and public sector bodies. Internal resources (R&D expenditure) and employment of technicians were found to be a much stronger determinant of product innovativeness.

This result is supported by earlier research. Spann & Souder (1991) examined the transfer of technology from US government facilities to start-up firms and concluded that in-house skills and efforts were by far the most important source of technical information. Rothwell and Dodgson (1991) also stress the importance of internal resources, 'the most important factors determining a SMEs' propensity and ability to access external sources of technology are *internal* to the firm' (p.131). This view would be supported by the work of Cohen & Levinthal (1989) who found that in addition to its obvious role of generating product innovations, R&D also increases a firms 'absorptive'[1] capacity. They suggested that in order for a firm to be able to exploit external knowledge it needs to have the internal skills to understand this knowledge and the uses to which it can be put.

Veugelers (1997) found that firms with a higher level of internal R&D expenditure are more likely to be involved in external R&D cooperation. This makes sense from the perspective of transactions costs. Firms with a good understanding, or the ability to easily develop a good understanding, of the technology that they are about to acquire (or collaborate in developing) will experience lower transaction costs than firms without this understanding, (Tidd, Bessant & Pavitt, 1998).
More recently Love and Roper (2001) in their study, that included 500 Irish firms, found that '…in none of the three countries does the intensity of plants' networking activity have any significant effect on either the extent or success of plants' innovation activity.' They found support for this result in the findings of Brouwer and Kleinknecht (1996) who, in a study of Dutch firms, found 'against all expectations … very little evidence that firms which collaborate on R&D or acquire external technological knowledge have a higher innovation output'. It should be noted however that Love and Roper limited their definition of networking to collaborative R&D; they did not include collaboration with customers or suppliers.

Love and Roper made two other interesting observations based on their data. First, they found that technology transfer and networking were used as substitutes for internal R&D rather than as complementary processes – this would help to explain the lack of evidence for increased R&D intensity linked with increased networking. Second, in common with earlier research by Acs and Audretsch (1988), they found that R&D intensity varied inversely with firm size but that there was little evidence for a relationship between product innovation success and firm size.

LaBahn et al. (1996) studied new product development in small (fewer than 100 employees) manufacturing firms. Their model examined the impact of non-technical outside assistance on NPD cycle time. They

---

[1] Absorptive capacity is defined, in this context, as the ability of a firm to absorb a new technology.

concluded that 'outsider involvement further complicates product development and delays the process'. However, they moderate their findings with reference to an earlier paper by Clark (1989), commenting that 'the benefits of non-technical outside assistance may be a function of the user's ability to integrate the expertise'. Again, internal skills are seen as more important than external interaction.

2.5.2 R&D/Marketing Integration

Many studies of NPD have stressed the importance of integration between the functions involved in the NPD process. (Shrivastava & Souder, 1987; Crawford, 1994). The degree to which marketing and R&D personnel collaborate in the development of new products has been shown to be linked with new product success by many authors (Griffin & Hauser, 1996; Song & Parry, 1997; Leenders & Wierenga, 2002).

Griffin & Hauser (1996) suggest six integrating mechanisms that may be of use in improving the R&D-marketing interface in NPD; co-locating R&D and marketing, moving staff between functions, informal social networks, integrating structures, joint incentive systems, and formal development processes. However, they do not test the impact of any of these mechanisms on the success of product development. Leenders & Wierenga (2002) take up the work of Griffin & Hauser and test the six proposed mechanisms for improving integration. They also test the importance of ICT (Information and Communication Technology).

Leenders & Wierenga found that many mechanisms that were positively linked with integration were not positively linked with success. Having formal management processes was the mechanism most strongly linked with improved integration but it had a negative impact on NPD performance. The authors suggest that this may be due to lower employee initiatives and reduced flexibility. Physical distance between R&D and marketing, equal incentives and cross-functional teams were also linked with increased R&D-marketing integration. The relationship between informal social systems and success was positive but not strong and the authors found that they contributed no additional benefits when cross-functional teams and ICT were already being used. ICT was the only factor that was found to have a significant positive impact *directly* on NPD performance. This is interesting as the extensive use of ICT is relatively recent and it is a factor that has not been considered in previous studies (Griffin & Hauser, 1996)

Song & Parry (1997) found that in both Japan and the US cross-functional integration was linked with project success. In particular they found that high levels of cross-functional integration led to; a better flow of market and customer knowledge across all project team members, closer R&D-customer contact during the later stages of NPD, and, faster project 'kill' decisions.

The studies discussed above assume that R&D-marketing integration is a good thing at all stages of product development, Song et al. (1998) questioned whether or not this was the case and found that having all functions involved at all stages could be counter productive. The explanation they offer for this is that conflicts are more likely to occur when all functions are involved, also the time and cost associated in getting all functions to meet and reach a consensus may outweigh the benefits.

The R&D-marketing interface is not considered to be a central issue in small firms. Small firms are viewed by many authors as having 'entrepreneurial management styles and structures which are characterised by informal control mechanisms, adaptability, flexibility and open communication channels' (Berry & Taggart, 1998), see also (Covin, 1990; Rothwell & Dodgson, 1991). This perception of small firms suggests that they should not have difficulty with communications between R&D and marketing. Souder (1995) found that in small firms movement between R&D and marketing is very free with R&D and marketing personnel exchanging roles. Rothwell & Dodgson (1991a) also saw small firms as having 'efficient internal communication' but being disadvantaged in terms of their external linkages.

In summary, there is a consensus that cooperation between R&D and marketing at most stages of NPD projects is linked with success. While several barriers to R&D-marketing integration have been identified in large firms, they have not been seen in small firms who have been found to have good internal communications.

2.5.3 Methods of Organising for NPD

Clark & Wheelwright (1992) described four types of organisational structures that are commonly used for NPD; (1) functional teams, (2) lightweight teams, (3) heavyweight teams and (4) dedicated teams. They concluded that heavyweight teams, with an emphasis on functional integration and strong leadership, are the most suitable for NPD. Cooper (1994) concurs, recommending that a good organisational design for developing new products should feature a dedicated, cross-functional team that is accountable for a project from beginning to end, led by a product champion and with strong top management support.

Small firms are assumed to enjoy efficient internal communications and involved management (Rothwell & Dodgson, 1991). Yap & Souder (1994) found that a non-participating, authorative management style correlated positively with project success. This illustrates that the impact of organisational factors on the success or failure of new product development must be considered in a very different way for small and large firms.

Lee et al. (2000) examined the impact of eight different methods of organising NPD projects on new product success. They compared NPD management in Korea and the US and found that in Korea the most popular method of organising for NPD was using a NPD department while in the US the most popular method was under a NPD manager – both of these methods are obviously quote similar. The only method that led to significant differences in performance was the use of a venture team. In Korea, projects that employed venture teams reported *lower* performance that those that did not while in the US projects that employed venture teams reported *higher* performance than those that did not.

2.6 Summary of Literature Review

This chapter has reviewed literature on the determinants of NPD success in three areas, Strategic and Market Factors, Development Process Factors and Organisational Factors. The review has not been restricted to publications addressing small firms; a more general approach has been adopted. The literature available on some of the topics included in this review is very extensive therefore for practical reasons this review cannot claim to be exhaustive. The focus of this study is on the management of NPD in small firms therefore the key points that emerge from the literature reviewed that addressed small firms are presented below.

2.6.1 Summary of Strategic and Market Factors

Several researchers have identified strategic and market factors that are linked with success in small firms. The key points that emerge from the literature are summarised below.

1. There appears to be a level of consensus in the literature reviewed that small firms should develop new products that are compatible with existing technical standards (Yap and Souder, 1994), that they should adopt one core technology and avoid high levels of diversification (Meyer & Roberts, 1986; Dodgson & Rothwell, 1991).
2. The literature reviewed offers conflicting views on what market characteristics are linked with success in small firms. It is not clear whether small firms should target new rapidly changing markets (Acs & Audretsch, 1988) or mature markets (Yap and Souder, 1994), or indeed whether market characteristics have an impact on new product success at all (Hoffman et al., 1998).
3. Marketing and technological synergy are linked with new product success within the literature surveyed (Song & Parry, 1999; Balbontin et al. 1999) but the impact of synergy on small firms has not been addressed to any great extent.
4. A lack or a shortage of resources for NPD has been identified by several authors as a barrier to firms undertaking product development (Fitzgerald & Breathnach, 1994; Roper et al., 1996; Hoffman et al., 1998). But there is no evidence of a link between the level of NPD resources and new product success.

2.6.2 Summary of Development Process Factors

The literature reviewed on development process factors did not provide much insight in to the specific needs of small firms – several authors have commented on the lack of research into the NPD process in small firms (Acs & Audretsch, 1988; Berry & Taggart, 1998; Hoffman et al., 1998; March-Chordà et al. 2002; Walsh et al. 2002). Some of the key points that emerged are listed below.

1. NPD processes in small firms tend to be informal (Roper, 1997; Hoffman et al., 1998; Santarelli & Sterlacchini, 1990). Some studies have linked more complete NPD processes with increased

success (Cooper & Kleinschmidt, 1986; Rochford & Rudelius, 1997) though no studies were found that linked NPD process formality (or completeness) with new product success in small firms.

2. Top management support and skill are important for new product success in small firms (Hoffman et al. 1998) though most NPD projects undertaken by small firms have the support of top management (Maidique, 1980).

3. Early studies established a link between marketing and technological skills and new product success (Cooper, 1987; Cooper & Kleinschmidt, 1987; Zirger & Maidique, 1990) though more recent studies are less conclusive and question whether marketing or technological skills are more important (Wren et al. 2000). No studies were found that examined the importance of marketing and technological skills to small firms.

4. Studies linking proficiency at different NPD process stages with new product success stress the importance of predevelopment activities (Cooper & Kleinschmidt, 1987), though these have concentrated on large firms. There is also evidence that the importance of the different NPD stages depends on the type of products being developed and the needs of the organisation (Bessant & Francis, 1997; Song & Montoya-Weiss, 1998; Loch, 2000).

5. There is general agreement that project managers skills are linked with new product success, though the way in which they are linked appears to vary with firm size and location (Yap & Souder, 1994; Yap et al., 1998; Souder & Song, 1998; Souder & Jenssen, 1999; Lee et al. 2000).

6. A range of methods to reduce NPD cycle time has been identified in the literature (Eisenhardt & Tabrizi, 1995; Zirger & Hartley, 1996; Griffin, 1997; Datar et al., 1997) but the effectiveness of these methods depends on the type of product being developed and on firm size. Also, the link between time to market and new product success is not clearly identified within the literature.

2.6.3 Summary of Organisational Factors

Small firms are very different from large firms in the way that they are organised. The organisational factors that have been linked with new product success in small firms are summarised.

1. A significant body of literature challenges the view that the involvement of external organisations in NPD projects increases success (LaBahn et al., 1996; Freel, 2003; Spann et al. 1991; Veugelers, 1997; Rothwell & Dodgson, 1991; Love & Roper, 2001). Results relating to both large and small firms suggest that internal capabilities are the most important factors in determining NPD success and that the effective use of external organisations in NPD projects requires a high level of internal skill.
2. Increased integration between R&D and marketing is linked with increased new product success in large firms and many authors have identified ways in which the R&D/marketing interface can be improved (Griffin & Hauser, 1996; Song & Parry, 1997; Leenders & Wierenga, 2002). However, R&D/marketing integration has not been identified as a key issue for small firms who are assumed to have good communication between all functions (Berry & Taggart, 1998; Rothwell & Dodgson, 1991a)
3. The literature examining methods of organising NPD is based mainly on large firms (Clark & Wheelwright, 1992). Organising for NPD is not considered to be a problem for small firms (Rothwell & Dodgson, 1991)

Table 2.4 summarised the literature in terms of the contribution of the different authors reviewed. This summary illustrates clearly that the management of NPD in small firms has not been researched to any great extend. Several gaps existing in the literature dealing with the NPD management in small firms these are summarised in Section 2.6.4.

Strategic and Market Factors

Determinants	Contributors	Comment
New Product Characteristics	Cooper, 1990 Karakaya and Kobu, 1994; Yap and Souder, 1994	Product advantage is one of the most critical success factors in large firms but is not always associated with success in small firms.
	Walsh et al., 2002; Santarelli & Sterlacchini, 1990; Bower & Christensen, 1995	Product innovation in small firms is mainly incremental. However large firms also have difficulties with disruptive technologies.
	Meyer and Roberts, 1986; Ali et al., 1995 Yap and Souder, 1994; Hoffman et al., 1998	Small firms are advised to avoid very radical new products and high levels of differentiation, to adopt one core technology and pursue cost-leadership or focus strategies.
Market Characteristics	Cooper, 1979; Cooper and Kleinschmidt, 1987; Acs and Audretsch 1988; Abernathy & Utterback, 1978	The extent to which market potential affects new product success in either small or large firms is still open to debate. However, research suggests that small firms have an advantage in newer markets while large firms have an advantage in capital-intensive markets.
	Porter, 1980; Yap & Souder, 1994; Hoffman et al., 1998; Yap and Souder, 1994	There is little consensus in the literature about the impact of market competitiveness in new product success in either small or large firms. However there is evidence to suggest that small firms can be more successful in mature markets.
Marketing and Technological Synergy	Maidique and Zirger, 1984; Cooper, 1994; Yap and Souder, 1994; Song & Parry, 1999; Balbontin et al. 1999	Technical and marketing synergy have been linked with NPD success in both large and small firms.
Resources for NPD	Acs & Audretsch, 1988; Page, 1993; Fitzgerald and Breathnach, 1994; Roper et al., 1998; Hoffman et al., 1998; Wakasugi & Koyata, 1997	The availability of resources for NPD is an issue for about a third of large and small firms. Though R&D intensity is much higher in small firms there is evidence that they are as efficient as large firms in developing new products.

Development Process Factors

Determinants	Contributors	Comment
Top Management Support	Maidique, 1980; Maidique & Zirger, 1984; Hoffman et al., 1998; Swink, 2000	Top management support is linked with success in large firms while top management skill is important in small firms.
Marketing and Technological Skills	Cooper and Kleinschmidt, 1987; Cooper, 1987; Zirger & Maidique, 1990; Wren et al., 2000	Marketing and technological skills are linked with new product success though there is no consensus as to which skills are the most important.
Development Process Proficiency	Cooper, 1979; Cooper and Kleinschmidt, 1986; Cooper and Kleinschmidt, 1987; Song & Montoya-Weiss, 1998	Proficiency of predevelopment and technical activities is linked with NPD success in both small and large firms.
	Cooper and Kleinschmidt, 1986; Santarelli & Sterlacchini, 1990; Roper, 1996; Rochford & Rudelius, 1997; Bessant & Francis, 1997 Song & Montoya-Weiss, 1998; Veryzer, 1998	A formal NPD process is necessary to effectively manage NPD projects. However, a considerable amount of R&D in small firms is informal and is less effective than the formal R&D undertaken by large firms. Also the degree of newness of a product can determine what type NPD process is most effective.
	Bessant & Francis, 1997; Loch, 2000	Firms should implement an NPD process that suits their requirements – there is no 'best' process.
Project Management Skills	Clark & Fujimoto, 1991; Wheelwright & Clark, 1992; Yap et al. 1998 Norrgren & Schaller, 1999	Project managers organisational authority and skill are linked with new product success.
	Yap & Souder, 1994; Souder & Song, 1998; Souder & Jenssen, 1999; Lee et al. 2000	The impact of Project Management Skills varies with firm national location and size.
Cycle-time Practices	LaBahn et al., 1996; Ali et al., 1995 Walsh et al., 2002; Dodgson & Rothwell, 1991	Small firms are able to get new products to market several times faster than large firms and can speed up their product development by developing incremental new products but outside assistance increases time-to-market.
	Eisenhardt & Tabrizi, 1995; Callahan & Moretton, 2001; Griffin 1997	The effectiveness of a particular technique in reducing time-to-market is dependent on the type of product (complexity, newness) being developed.

	Contributors	Comment
	Eisenhardt & Tabrizi, 1995; Zirger & Hartlet, 1996; Griffin 1997; Datar et al. 1997;	Techniques that have been found to be effective in reducing NPD cycle times are: cross-functional teams, CAD/CAM, project leader power, frequent prototypes and testing, project goals and process formality.

Organisational Factors

Determinants	Contributors	Comment
Involvement of External Organisations	Cohen & Levinthal, 1989; Spann & Souder, 1991; Freel, 2003; Love & Roper, 2001; Brouwer & Klienknecht, 1996; LaBahn, 1996	Internal resources are more important than external linkages to new product success. Many authors argue that external interaction on NPD projects is not linked with success.
	Freel, 2003; Love & Roper, 2001	There is not a strong link between networking and the level of NPD in small firms.
R&D/Marketing Integrating	Guptu et al. 1987; Griffin & Hauser, 1996; Leenders & Wierenga, 2002; Song & Parry, 1997 Berry & Taggert, 1998; Rothwell & Dodgson, 1991; Song et al. 1998	A strong link has been found between effective R&D-marketing communication and NPD success – though this is not seen as a central issue for small firms. Techniques useful in improving R&D-marketing communications include co-location, culture, reward systems and ICT.
	Walsh et al., 2002; Dodgson and Rothwell, 1991; Acs & Audretsch, 1988	Small firms have several advantages over large firms, they are more flexible, can develop new products in less time.
Methods of Organising NPD	Clark & Wheelwright, 1992; Cooper, 1994	Functional integration and strong leadership are linked with new product success.
	Covin, 1990; Rothwell & Dodgson, 1991; Berry & Taggart, 1998	Small firm have better internal communications, more flexibility, and more entrepreneurial management than large firms
	Rothwell & Dodgson, 1991; Yap & Souder, 1994; Lee et al. 2000	The most effective method of organising NPD projects depends on firm size and location.

Table 2.4 Summary of contributions on determinants of NPD success

2.6.3 Knowledge and Gaps

Within the body of literature reviewed in this chapter there are areas where there is some consensus about how NPD in small firms should be managed, but there are many more areas where research into NPD management in small firms is lacking. The perceived knowledge about NPD in small firms is summarised below.

1. Small firms should develop compatible products, adopt one core technology and avoid diversification.
2. Top management support and project management skills are linked with increased new product success in small firms.
3. NPD processes in small firms are informal, the relationships between R&D and marketing are generally good and the way in which small firms organise NPD is not regarded as a problem.
4. Recent publications question the general perception that involving external organisations in NPD projects is beneficial and also suggest that managing external involvement can be difficult and time consuming for small firms.

The following gaps in the literature about the management of NPD in small firms have been identified.

1. No studies were found that examined the link between technical and marketing skills and synergy and new product success in small firms.
2. There is no consensus in the literature about which market strategies are linked with new product success in small firms.
3. While a lack of resources has been identified as a barrier to product innovation in small firms, the link between NPD resources and new product success has not been examined.
4. The literature suggests that the importance of different NPD stages depends on the context within which a firm operates but no studies were identified that examined the importance of different NPD stages in small firms.
5. The effectiveness of different NPD cycle time reduction strategies depends on the technology being developed and the size of the firm but there is no clear indication in the literature of which cycle time reduction strategies are useful in small firms. Nor is the link between NPD cycle time and new product success established for small firms.

The objective of this study can be achieved by examining the extent to which the perceived knowledge about NPD in small firms applies to small Irish electronics firms and also by trying to fill in some of the gaps identified above. This can be shown by extending the illustration of the research question presented in Figure 2.1 to include the NPD management factors addressed in this review, this is done in Figure 2.6.

Within this review one particular stream of publications have proven to be comprehensive in its analysis of the factors that distinguish between new product success and failure. The INTERPROD study has included data from many different countries and has resulted in over ten publications. Also the models presented by several of the researchers involved in this study, see Figure 2.4, appear to be useful in answering the research question posed at the start of this chapter. Therefore, in an attempt to develop a deeper understanding of the management practices in small and large electronics firms, and the way in which these management practices are related to new product success the results of the INTERPROD study will be examined in more detail.

2.7 Conclusion

This chapter has presented a review of the determinants of new product success highlighting references to small firms. The review has been structured around three topic areas 1) strategic and market factors, 2) development process factors, and 3) organisational factors; covering several factors within each area. The literature has shown that there are differences between the way in which NPD is managed in small and large firms and has also highlighted the different success factors that apply to large and small firms.

Chapter 3 presents a detailed review of the publications that have resulted from the INTERPROD research project. Based on the INTERPROD and the other publications reviewed in this chapter a research model will be developed in Chapter 4.

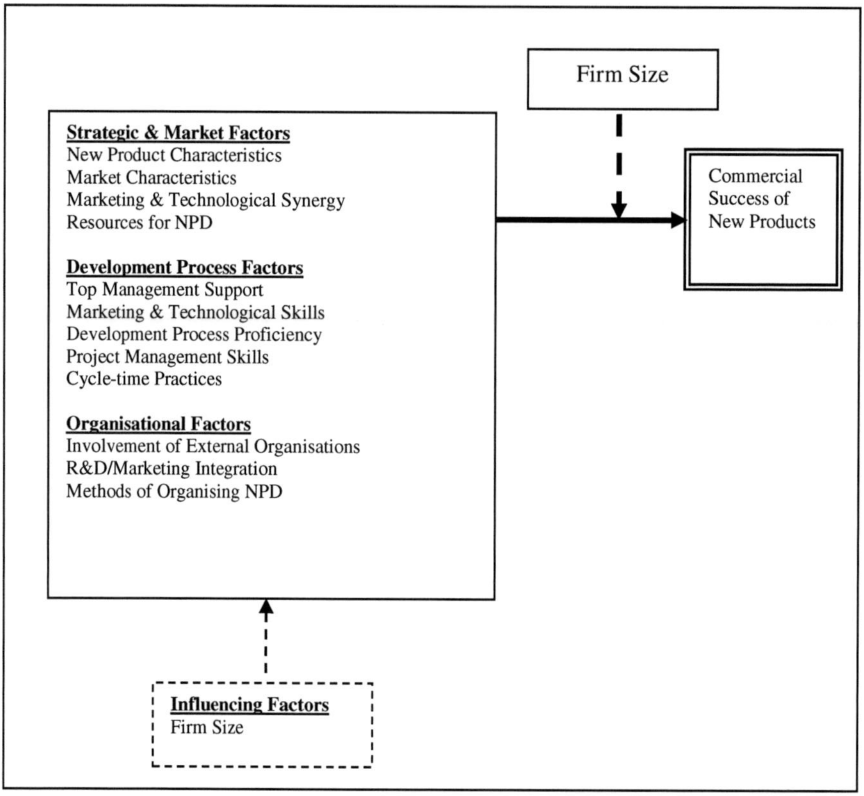

Figure 2.6 Proposed Research Model

CHAPTER THREE

THE INTERPROD STUDY

3.1 Introduction

This chapter provides an overview of the INTERPROD project and a review of the literature that has resulted from this project. The chapter starts by giving a description of INTERPROD and how it was structured and conducted. In Section 3.2 the 13 papers that have been published based on INTERPROD are reviewed in detail. Section 3.3 contains a summary of the INTERPROD publications and draws out any significant themes that emerge from the data. An overall model of the INTERPROD study, positioning each of the published papers, is also presented in Section 3.3. Section 3.4 contains the conclusion of this chapter.

3.2 INTERPROD

INTERPROD is an international study of the factors that influence the success or failure of new products developed by high technology firms. It was set up in 1994 by Prof William Souder at the University of Alabama in Huntsville, USA, and was funded by the US National Science Foundation. In total, fifteen countries were involved in the study, these are listed in Table 3.1. To date the INTERPROD project has resulted in 13 journal articles, (these will be discussed in this chapter) and many conferences papers and reports.

INTERPROD focused on four industries: telecommunications, electronics, biotechnology and pharmaceutical. Firms that produced industry products (i.e. sold to other firms) rather than consumer products were included in the study. The project was coordinated through the Centre for the Management of Science and Technology (CMOST) in Alabama. All data, once collected, were sent to CMOST where they were coded and sent back to the country Team Leader in a soft format. There was a certain amount of central analysis at CMOST but most of the publications resulted from individual cooperation between the participants. Firm and project names were disguised but firms were identified by their SIC codes.

Country	Team Leader	Number of Firms	Number of Projects
US	**Prof Wm Souder**, University of Alabama in Huntsville	39	113
Australia	**Dr Robert Mellor**, University of Western Sydney	25	47
Japan	**Prof Michael Song**, University of Tennessee	17	65
New Zealand	**Dr David Buisson**, University of Otago	29	74
Holland	**Prof Rudy Moeneart**, Free University of Brussels	70	96
France	**Prof Thierry Gonard**, Nantes Graduate School of Business	9	16
Norway	**Prof S.A. Jenssen**, Norland Research Institute	30	54
Belgium	**Prof Rudy Moeneart**, Free University of Brussels	38	66
Singapore	**Dr Chee Meng Yap**, National University of Singapore	16	30
Korea	**Dr Jinjoo Lee**, Korea Advanced Institute of Science and Technology	15	52
Taiwan	**Dr Her-Juin Sheu**, National Chiao-Tung University	30	30
Germany	**Dr Helmut Schmalen**, University of Passau	10	40
Sweden	**Prof S.A. Jenssen**, Norland Research Institute	9	18
UK	**Prof Rachel Davis Cooper**, Salford College of Technology	63	102
Total:		**400**	**803**

Table 3.1 Summary of International INTERPROD Participation

The INTERPROD survey gathered data using a series of nine questionnaires that were completed by different members of the participating firm. Each questionnaire was designed to take no more than 5-10 minutes to complete. Most of the questions simply required the respondent to select an answer from a number of options or from a Likert scale. Copies of the questionnaires are given in Appendix 1.

The questionnaires were numbered INT-A to INT-P and were aimed either at a company or a project level. INT-A, INT-F, INT-G, INT-L and INT-P were used to collect general information on the company and the environment within which it operates. While INT-B to INT-E were aimed at specific new product projects within the company.

Each company participating in the research was asked to provide data on two new product successes and two failures. One successful product should have been launched under conditions of market certainty and one into an uncertain market, failed products were also to be divided between certain and uncertain markets. However, this was not always practical and some firms only reported on one or two projects. Table 3.2 gives a summary of each questionnaire. Respondents varied for each type of questionnaire but included the MD, technical manager/director, marketing manager and R&D engineers.

The manner in which the INTERPROD data were collected varied from country to country. In the US researchers worked closely with the firms involved, the ratio of projects to firms in the US is quite high with each firm reporting on an average of 3 projects. In Japan also the ratio of projects to firms was high. In the UK the questionnaires were sent by post and though the number of firms involved was high, 63, the number of projects per firm was very low. Additionally, the UK data are not balanced in terms of market certainty.

The overall framework upon which INTERPROD is based is shown in Figure 3.1 and the thirteen papers published to date are listed in Table 3.3. Question codes have been added where possible to allow the reader to locate questions in Appendix 1. These have been placed in square brackets after each question. The thirteen INTERPROD papers will be analysed in sections 3.2.1 to 3.2.13, within these sections the publications summarised in Table 3.3 are reviewed in chronological order.

The analysis of the Irish INTERPROD data presented in this thesis has resulted in two publications (Ledwith, 2000a; Ledwith 2000b), these are presented in Appendix 2. However, in order to present a logical development of the arguments these will not be discussed at this stage but will be referred to in later chapters.

Questionnaire	Completed by	Unit of Analysis	Subject
INT-A	CEO, Managing Director	Company	General overview of the company
INT-B	Project Leader/Manager (Technical or Marketing)	Project	Management of new product development
INT-C	Project Leader/Manager (Marketing)	Project	Market Environment
INT-D	Project Leader/Manager (Technical)	Project	Technical Environment
INT-E	Project Leader/Manager (Technical)	Project	Level of innovation
INT-F	R&D Manager/Director	Company	Cycle time of new products
INT-G	Representative of company culture	Company	Company culture
INT-L	R&D or Marketing Manager with direct responsibility for new product projects	Company	Organisational learning
INT-P	R&D Manager/Director	Company	Government Policy

Table 3.2 INTERPROD Questionnaires

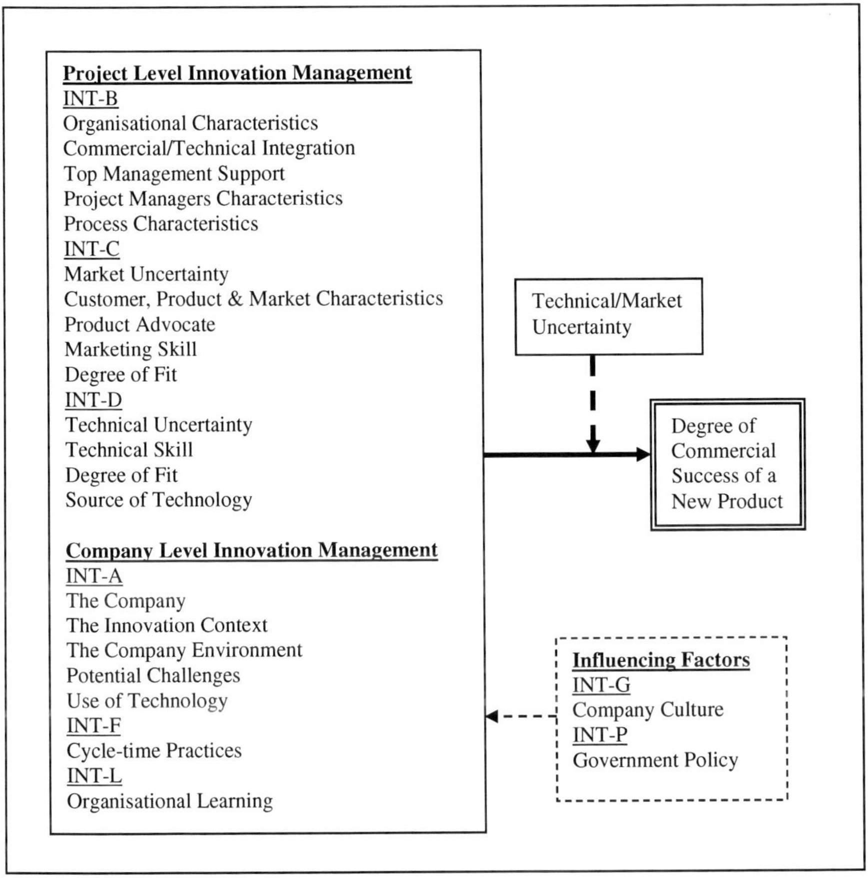

Figure 3.1 Overall Framework for INTERPROD

Authors, year and journal	Countries Involved	Focus of Paper
1. Yap & Souder, 1994, JPIM.	US	Factors influencing success in small high-technology firms.
2. Souder & Song, 1997, JPIM.	US and Japan	Impact of product design and market strategies on success in small and large US, and large Japanese firms under conditions of high and low market uncertainty.
3. Song, Souder & Dyer, 1997, JPIM.	Japan	Skills, Synergy and design sensitivity
4. Souder, Buisson & Garret, 1997, JPIM.	US and New Zealand	Customer driven NPD
5. Gupta & Souder, 1998, Research Technology Management.	US	Comparison of management practices in companies reporting short and long NPD cycle times.
6. Souder, Sherman & Davis-Cooper, 1998, JPIM.	US and UK	Impact of R&D/marketing and R&D/customer integration on NPD effectiveness under conditions of high and low market uncertainty.
7. Souder, Song & Kawamura, 1998, Research Technology Management.	US and Japan	Controlled Scrimping
8. Souder & Song, 1998, JPIM.	US and Japan	Management practice in Japan and US
9. Balbontin, Yazdani, Cooper & Souder, 1999, International Journal of Technology Management.	US and UK	Success Factors
10. Souder & Jenssen, 1999, JPIM.	US and Scandinavia	Management practice
11. Lee, Lee & Souder, 2000, Technovation.	US and Korea	Organisation Characteristics
12. Wren, Souder & Berkowitz, 2000, International Marketing Management.	US, New Zealand, Korea, Scandinavia and Belgium	Market orientation
13. Sherman, Souder & Jenssen, 2000, JPIM.	US and Scandinavia	Cross functional Integration and NPD cycle time

JPIM = Journal of Product Innovation Management

Table 3.3 INTERPROD Publications

3.2.1 Entry Strategies: Yap & Souder, 1994

The first INTERPROD paper published was Yap and Souder (1994). Chee Meng Yap developed and administered five of the questionnaires (INT-A to INT-E) as part of his PhD (Yap, 1992). He explored the impact of technical and market uncertainty on new product success in small entrepreneurial, electronics firms. The firms he studied were small (fewer than 500 employees and with annual sales below US$75 million), high-technology, electronics firms located in Huntsville, Alabama, US. In his PhD, Yap came to two major conclusions. The first was that the factors that affected the success or failure of NPD in the firms he studied were not the same as those reported in the literature for Fortune 500 firms or for start-up firms. The second conclusion was that the level of technical and market uncertainty had an impact on the practices and strategies that led to commercial success in the firms he studied. His key findings were published in 1994 (Yap & Souder, 1994).

Yap studied 48 projects from 12 firms, the projects were equally spilt between success and failure and between high and low technical and market uncertainty. Successful projects were those that exceeded the expectations of the organisation while failures were those that failed to meet expectations. This is illustrated in Table 3.4. An attempt was made to avoid using products that scored 3, i.e. met expectations.

Score	Commercial outcome of the product	Category Assignment
1	Far below our expectations	Failure
2	Slightly below our expectations	Failure
3	Consistent with our expectations	Null
4	Slightly above our expectations	Success
5	Far above our expectations	Success

Table 3.4 Project outcome scoring system

Market uncertainty was measured using responses to three statements.

1. Our company was familiar with the market. [C5]
2. The customers' needs were well defined. [C6]
3. The customers' needs could be readily translated into product specifications. [C7]

Respondents who agreed with two or more of the above statements were deemed to be operating under low market uncertainty. While those who disagreed with two or more of the statements were seen to be operating under conditions of high market uncertainty.

Technical uncertainty was measured using another three statements.

1. The technological phenomena on this project were well known by us. [D6]
2. When a technical problem came up in this project, we had a storehouse of information to draw from. [D8]
3. The technical aspects of this product development process involve complex feedback, trial and error methods. [D7]

The last of these three measures was inverted as a positive response to this question indicates a high level of technical uncertainty in contrast to the other two measures where a positive answer indicates low uncertainty. Again, agreeing with two or more of these statements indicated low technical uncertainty, otherwise high uncertainty was assumed.

Yap & Souder examined 35 items that were clustered under 7 factor headings: 1) Project Synergies, 2) Product Characteristics, 3) Skills Level, 4) Market Characteristics, 5) Technology Sources, 6) Entry Strategy and 7) Organisational Characteristics. Spearman correlation coefficients were computed between the five-point commercial outcome scales and the scores for each of the 35 factors. Results for 5 of these 7 factors were found to be inconsistent with prior findings; the correlation coefficients for these 5 factors are included in Yap & Souder's paper and are summarised in Table 3.5.

	Correlation Coefficients between factor scores and project outcomes			
	Market Uncertainty		Technical Uncertainty	
	Low	High	Low	High
Technology Sources:				
Hiring key personnel from another	-	$-.58^a$	-	$-.36^d$
organisation [D17]	$.66^a$	NS	$.75^a$	NS
Licensing [D21]	$.69^a$	NS	$.72^a$	NS
Developer's own previous contract research	-	NS	$-.31^e$	$-.49^c$
[D19]	$.70^a$	NS	NS	$-.79^a$
University research [D23]	NS		NS	
Degree of own off-the-shelf technology used	NS			
[D25]				
Organisational Characteristics:				
Degree of non-participative style of project	$.83^a$	$.52^b$	$.84^a$	NS
manager [B31]	NS	NS	$.50^c$	$-.68^a$
Degree to which the organisation is organic				
[B16]				
Product Characteristics:				
Performance superiority as a product	-	NS	NS	NS
advantage [C38]	$.76^a$	NS	NS	NS
Unique features as product advantage [C40]	-	$.72^a$	NS	$.61^a$
Compatibility as product advantage [C46]	$.67^a$			
	NS			
Market Characteristics:				
Degree of market maturity [C15]	$.65^a$	$.81^a$	$.40^d$	$.52^c$
Degree of importance of decision [C10]	NS	NS	NS	$-.63^a$
Entry Strategies:				
Positioning: Against another product [C28]	$.49^c$	NS	$.41^c$	NS
Positioning: Product class dissociation [C29]	NS	$-.77^a$	-	NS
Positioning: Benefits and solutions	NS	$-.74^a$	$.57^a$	$-.70^a$
emphasised [C25]	$.61^a$	NS	NS	NS
Distribution: Developers own salesforce [C30]	$.33^e$	$.56^b$	$.53^b$	$.62^s$
Distribution: Manufacturers representatives			NS	
[C32]				

Correlation coefficients statistically significant at: $^a p \leq .005$; $^b p \leq .01$; $^c p \leq .05$; $^d p \leq .10$; $^e p \leq .20$;
All are two-tailed tests with N = 24.

Table 3.5 Summary results from Yap & Souder, 1994

Table 3.6 summarises the overall findings of this paper.

Findings consistent with prior studies	**Findings inconsistent with prior studies**
• High levels of project synergies (engineering, marketing and production skills and product line) were found to correlate with success. • High levels of skills (engineering, marketing, sales and marketing, and project management) were found to correlate positively with success. • The integration of different departments on NPD projects and a high flow of information between the groups involved were linked with success. • Early top management involvement and an influential project manager was linked with success. • A cost effective product with good customer support and service was more likely to be successful. • The number of competitors in the market was negatively correlated with success while high growth potential markets were positively correlated.	• Hiring essential personnel from other organisations was found to hinder success. • Licensing was only an effective source of technology under low technical and market uncertainty. Acquiring technology from university research was not linked with success. • A non-participative project management style was correlated with success in all except highly technical uncertain projects. • Superior products and unique features were not linked with success except under conditions of low market uncertainty where they were found to hinder commercial success. • Mature markets were found to foster project success. • Positioning a product against another product and using the developer's own sales force promoted project success. • Positioning products based on uniqueness (class dissociation) or based on benefits and solutions hindered success especially in markets characterised by high uncertainty.

Table 3.6 Findings of Yap & Souder (1994)

Yap & Souder summarise by making a series of recommendations to small high-technology firms. These are:

1. Avoid hiring external experts as a form of technology acquisition.
2. Licensing is only a good idea under conditions of low market or technical uncertainty.
3. Technologies developed through government contracts or off-the-shelf technologies have a low likelihood of success.
4. Project managers should have good technical and market knowledge and should be given broad authority to make decisions.
5. Under conditions of high uncertainty offer compatible products.
6. Target a niche in a mature market.
7. Under conditions of low uncertainty firms should offer products that can be positioned against competing products.
8. Under conditions of high uncertainty in-house sales force should be used, under conditions of low uncertainty manufacturers representatives should be used.

3.2.2 Design and Market Parameters: Souder & Song, 1997

Souder and Song (1997) used the INTERPROD data to compare three different groups of firms, 1) small[1] US suppliers of electronics components, 2) Fortune 500 producers of electronics final products and 3) Japanese producers of electronics final products. They also took into consideration the level of uncertainty of the market into which a new product was launched, using the same measure developed by Yap (Yap & Souder, 1994). Their sample included 52 new products from 20 small US firms, 52 new products from 15 large US firms and 156 new products from 50 Japanese firms. These were spread evenly between successful and failed products under conditions of high and low market uncertainty. They examined a total of seven parameters each of which was constructed from 2 – 4 questions on the INTERPROD questionnaires, these are detailed in Table 3.7

[1] Note: Small here is defined as fewer than 500 employees and annual sales less than $200 million.

Design Parameters:	Market Parameters
Performance Superiority Positioned on specific product features [C24] Positioned on the benefits and solutions [C25] Superior Performance as a selling point [C38] **Technical Superiority** Technical Superiority as a selling point [C39] Product was at the forefront of technology.[b] **Radicalness** Unique Features were a selling point [C40] Positioned on product class dissociation [C29] Users required to change way of doing things. [C22] **Compatibility** Product Compatibility as a selling point [C46] Positioned on specific usage occasions. [C26][a] Product was suitable for many applications.[b]	**Users Eagerness to Adopt** Customer had a great need for product. [C8] Customers satisfied with existing products. [C12][a] Purchase of this product a major decision [C10][a] **Intensity of Competition** The market for this product competitive. [C16] There were many competitors in market. [C17] **Market Growth Potential** Market for this product growing quickly. [C14] The window of opportunity was big. [C18] The users' needs change quickly in market. [C9] The market for this product was mature [C15][a]

[a] Because of the 'negative' nature if these items compared to the parameter derived, their scales are reversed for statistical calculations.
[b] The author could not identify corresponding questions within the questionnaires provided by Prof Souder.

Table 3.7 Factors analysed by Souder & Song

The means of the responses to the INTERPROD questions (these were measured using Likert-type scales, 5 = Strongly Agree to 1 = Strongly Disagree) making up each parameter were calculated to determine the score for each parameter. These were then correlated against the outcome of the product. This was measured using the scale from 1 to 5 as shown in Table 3.4. Souder and Song excluded any projects that scored a 3 and considered those that scored 4 or 5 as being successful and those that scored 1 or 2 as being failures. The results of these correlations are shown in Table 3.8.

	Spearman Correlation Coefficients					
	Small US Firms		**Fortune 500 Firms**		**Japanese Firms**	
	Market Uncertainty		**Market Uncertainty**		**Market Uncertainty**	
Parameter	**Low**	**High**	**Low**	**High**	**Low**	**High**
Design						
Parameters:	$-.68^{a}$	NS	NS	$.68^{a}$	NS	$.81^{a}$
Performance	$-.73^{a}$	NS	NS	$.90^{a}$	NS	$.88^{a}$
Superiority	$-.62^{a}$	NS	$-.41^{c}$	$.81^{a}$	$-.40^{c}$	$.63^{a}$
Technical	$.79^{a}$	$.61^{b}$	$.43^{c}$	NS	$.48^{c}$	NS
Superiority						
Radicalness						
Compatibility						
Market						
Parameters	$.63^{a}$	$.51^{b}$	$-.63^{a}$	NS	NS	NS
Intensity of	$-.54^{b}$	$.48^{c}$	NS	$.50^{c}$	$.53^{b}$	$.62^{a}$
Competition	$-.47^{c}$	$.65^{a}$	NS	$.41^{d}$	$.51^{b}$	$.81^{a}$
Market Growth						
Potential Users						
Eagerness to Adopt						

Statistically significant at: $^{a}p \leq .001$; $^{b}p \leq .01$; $^{c}p \leq .02$; $^{d}p \leq .05$; N = 26, Two-tailed.

Table 3.8 Summary of results from Souder & Song, 1997

Their data show that firms should adopt different market and product strategies based on the level of market uncertainty. Additionally strategies that correlate positively with success in large firms in both countries are not significant or correlate negatively with success in small firms. They summarise their findings by recommending different strategies based on firm size, location and on market uncertainty, these are shown in Table 3.9.

Small US Firms	Fortune 500 US Firms	Japanese Firms
Low Market Uncertainty Conditions:		
• Design compatible products. • Select competitive markets	• Design compatible products.	• Design compatible products. • Select high growth markets • Select users who are eager to adopt
High Market Uncertainty Conditions:		
• Design compatible products. • Select competitive markets • Select high growth markets • Select users who are eager to adopt	• Design superior performance products. • Design superior technology products. • Design radical products • Select high growth markets • Select users who are eager to adopt	• Design superior performance products. • Design superior technology products. • Design radical products • Select high growth markets • Select users who are eager to adopt

Table 3.9 Effective Strategies suggested by Souder & Song (1997)

Table 3.9 suggests that the strategy adopted is more dependent on firm size than on location. Souder and Song also found that small firms should avoid many of the strategies that are associated with success in large firms, for example targeting high-growth markets, under conditions of low market uncertainty.

3.2.3 NPD Skills and Proficiency: Song, Souder & Dyer, 1997

Song et al. (1997) examined practices in 17 large, multi-divisional Japanese firms looking at 65 projects, 34 successes and 31 failures. The firms ranged in size from 50,000 to 180,000 employees with between US$10 billion and US$50 billion in annual sales. All firms operated in high-technology industries. They examined the impact of 5 variables on technical and marketing proficiency, product quality and new product performance. A conceptual model was developed, see Figure 3.2, linking these variables and the relationships indicated by solid lines in Figure 3.2 were tested. New product performance was measured using four items:

1. The commercial outcomes met our expectations. [B5]
2. The technical outcomes met our expectations. [B6]
3. There was a close fit between this product and our existing markets. [C52]
4. The customers were satisfied with the final product. (Not part of the Irish data.)

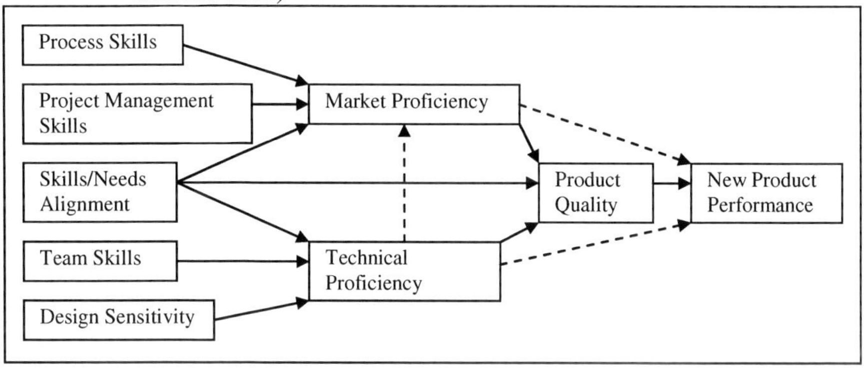

Figure 3.2 Conceptual Model proposed by Song et al., 1997

Each of the variables in the conceptual model was calculated as the average of several INTERPROD items. For example, the two proficiency variables were each calculated as the average of the measured proficiency at four of the eight stages in the proposed NPD process, see Table 3.10.

Technical Proficiency	Marketing Proficiency
Prototype Development activities [B52]	Exploratory activities [B50]
Prototype Testing activities [B53]	Concept Development activities [B51]
Manufacturing Start-Up activities [B55]	Market Development activities [B54]
Technical Service activities [B57]	Marketing Start-Up activities [B56]

Table 3.10 Calculation of Technical and Marketing Proficiency, Song et al. (1997)

Each of the relationships shown by solid arrows in Figures 3.2 was tested and the results are shown in Table 3.11. Marketing proficiency was found to be determined by process skills, project management skills and skills/needs alignment as expected. However, technical proficiency was not linked with skills/needs alignment; Song et al. suggest that this non-intuitive result may be due to the measure being used. The other unexpected result is the lack of a relationship between technical proficiency and product quality, again the scale used is questioned, in particular the fact that the exploration and concept development are not included in the measure of technical proficiency.

Variable	Marketing Proficiency	Technical Proficiency	Product Quality	New Product Performance
Process Skills [C48], [C49], [E10], [D11], [D12]	0.25[a]	-	0.08[a]	0.08[a]
Project Management Skills [B22], [B23], [B24], [B28], [B29], [B31], [B32], [B33]	0.43[a]	-	0.13[a]	0.14[a]
Skills/needs Alignment [C54], [D13], [D14], [D15]	0.38[a]	NS	0.73[a]	0.89[a]
Team Skills [B14], [B15], [B16], [B18], [B20]	-	0.25[a]	NS	NS
Design Sensitivity [C8], [C11], [C21], [C22][a], [C23]	-	0.82[a]	NS	NS
Marketing Proficiency See Table 3.10	-	-	0.30[a]	0.33[a]
Technical Proficiency See Table 3.10	-	-	NS	NS
Product Quality C36], [C25], [C37], [C41], [C45], [C46]	-	-	-	1.07[a]

A dash indicates that this relationship was not tested; [a] P<001; NS indicates nonsignificant at $\alpha = 0.01$

Note: for brevity the numbers of the INTERPROD questions used to make up each of the above variables are listed but the actual questions are not included – they can be found in Appendix 1.

[a]Item score reversed

Table 3.11 Relationships measured by Song et al. 1997

In summary, this paper applied path analysis to the Japanese INTERPROD data and found that new product success was linked to marketing proficiency and product quality but not technical proficiency. In turn, marketing proficiency is linked to process and project management skills and skills/needs alignment. Unexpectedly technical proficiency was not found to be a predictor of product success.

3.2.4 Technical and Market Proficiency: Souder, Buisson & Garret, 1997

David Buisson and Tony Garrett collected the INTERPROD data in New Zealand (NZ); this study is of particular interest to the Irish situation because of the similarities between the Irish and New Zealand economies and also because of its focus on small firms. The paper published evaluated the impact of 6 variables on new product success. In common with previous papers these variables were calculated as the means of component items. The variables examined were:

1. NPD Marketing Proficiency
2. Technical Skills
3. Marketing Skills
4. Project Manager Characteristics
5. R&D/Marketing Interaction
6. NPD Technical Proficiency

The items that were used to compute these variables were not the same as those used in previous papers. For example, NPD Marketing Proficiency and NPD Technical Proficiency were calculated using the measures shown in Table 3.12, this is obviously different from the composition of these variables presented in Song et al. (1997).

The inclusion of exploratory and concept development activities in the above measure of technical proficiency would be expected to overcome some of the limitations of the previous paper. However, the variation in calculating the different variables in the study means that results from both papers are not directly comparable. Another critical difference between these two papers is the way in which product success (or new product performance) is measured. Song et al. (1997) build a performance variable from the averages of four other measures all of which are project level measures while Souder et al. (1997) build a performance variable from a different four variables three of which are company level measures. Finally, when defining success and failure of products this paper includes projects that score 3 or higher on the scale presented in Table 3.4 while pervious researchers, notably Yap & Souder (1994) avoided projects that scored 3 (met expectations) and considered scores of 1 or 2 to be failures and scores of 4 or 5 to be successes.

Technical Proficiency	Marketing Proficiency
1. Exploratory activities [B50]	1. Market Development activities [B54]
2. Concept Development activities [B51]	2. Market Research [C48]
3. Prototype Development activities [B52]	3. Marketing Start-Up activities [B56]
4. Prototype Testing activities [B53]	4. Sales and Marketing activities [C49]
5. Manufacturing Start-Up activities [B55]	5. Launch activities[a]
6. Technical Service activities [B57]	

[a] This item was not recorded in the Irish data.

Table 3.12 Calculation of Technical and Marketing Proficiency, Souder et al. (1997)

The firms involved in this study all operated in high-technology sectors described by the following SIC two-digit codes: 28, 35, 36, or 38. They were all described as small and successful, meaning that they met Yap's (1992) definition of small, see above, and had marketed at least three successful new products in the past three years. The sample was balanced in terms of market uncertainty (as described in section 3.3.1) and numbers of successful and failed products. The final sample contained 75 US new products (38 successes and 37 failures) and 75 NZ new products (43 successes and 32 failures). Approximately half of each group of products (US success etc.) were released into familiar (certain) markets and half were released into unfamiliar (uncertain) markets. The overall sample was found to be statistically balanced.

The analysis in this paper focuses in the differences between the two samples. The means of the six variables were calculated for each country, as were the correlation coefficients between the variables and product performance. A summary of these results is shown in Table 3.13, for clarity the numbers of the INTERPROD variables used are listed.

For both countries the most significant measure is NPD marketing proficiency. When the means of the two sets of data are examined it is seen that the marketing proficiency is much higher in the N.Z. companies while project manager's skills are higher in the U.S. companies. The authors explain the lack of a significant relationship between technical skills and product performance as being due to the fact that all projects in both countries had high levels of technical skills therefore it was not found to be a differentiator between success and failure.

	Correlation coefficient between measures and product performance		**Mean score for each measure**	
Measure	**U.S.**[a]	**N.Z.**[a]	**U.S.**[a]	**N.Z.**[a]
NPD Marketing Proficiency	.537[b]	.671[b]	2.93[f]	4.03[f]
See Table 3.12				
Technical Skills	NS	NS	3.99	4.02
[D10], [D11], [D12], [E9]				
Marketing Skills	.335[c]	.573[b]	2.78[g]	3.39[g]
[C50], [C51], [C54]				
Project Manager	.511[b]	.309[d]	4.01[f]	3.02[f]
Characteristics				
[B28 – B32]	.453[b]	.465[b]	3.49	3.65
R&D/Marketing				
Integration	.382[c]	.488[b]	3.05[f]	3.82[f]
[B18], [B19], [B21], [B22]				
NPD Technical Proficiency				
See Table 3.12				

[a] N = 75 for each country:
[b] Significant at $p \leq .001$: [c] Significant at $p \leq .01$: [d] Significant at $p \leq .05$
[e] Means of a 5-point Likert scale (1 = Very low .... 5 = Very High)
[f] Significant level of difference, one tailed t-test <.001
[g] Significant level of difference, one tailed t-test <.01

Table 3.13 Summary results of Souder et al. (1997)

When the overall success of NPD in both sets of firms is examined it is found that the N.Z. firms score higher on every measure. Additionally, the ability to understand customer needs and to translate these needs into product specifications was found to be significant for N.Z. firms but not for U.S. firms. In discussing these results the authors conclude that the success of the N.Z. firms is due to their closer relationships with their customers and their ability to translate customer needs into products. They go on to discuss many of the features of the N.Z. firms; smaller than the U.S. sample, flatter, more organic organisations, better internal communications, greater overlap between the technical and marketing roles and more focussed on export than the U.S. firms. Many of these characteristics can also be found in the Irish firms who participated in the INTERPROD study.

3.2.5 Short Cycle-Time Firms: Gupta & Souder, 1998

Gupta and Souder (1998) used the U.S. INTERPROD data to examine the characteristics of firms who reported short cycle times. They analysed 110 projects in 38 U.S. companies, 16 small and 22 large, all operating in high-technology sectors. This is the same sample used in previous publications. They noted that within this sample about half the companies (the exact number is not given) reported shorter life-cycles than the industry average. The data revealed that these firms reported cycle-times over 40% less than the median value for the total data set, 38 companies. They also found that these 'short-cycle-time' companies generated more income from new products (those launched during the previous three years), and had higher profits, return on equity and new product success rates than others in the sample.

Further analysis of the data revealed five areas in which there were major differences between how the short- and long-cycle-time companies operated. These differences were:

1. The extent of user involvement in the NPD process
2. Management of NPD teams
3. Supplier involvement
4. Characteristics of the design process
5. The transfer of learning from one NPD project to another.

As in pervious studies each of the above areas was examined using data from several INTERPROD questions. These results are summarised in Table 3.14.

Area of analysis	Short-cycle-time Firms[a]	Long-cycle-time Firms	Sig.[b]
The extent of user involvement in the NPD			
process:	3.62	2.44	0.025
Frequency of contact with users is high [F7]	4.44	3.56	0.050
Users define and clarify needs early [F6]	4.69	3.44	0.005
Users try the product as it is developed [F8]	4.62	3.56	0.011
Users are given working prototypes early [F11]			
Management of NPD teams:			
Teams are used for product development [F10]	4.31	3.67	0.180
Teams have multiple people with leadership skills	3.31	2.00	0.035
[F17]	4.19	2.89	0.014
Team members choose their own roles [F18]			
Supplier involvement:			
Supplier personnel on development teams [F12]	3.94	2.22	0.002
Partnering with component supplier F23]	4.25	2.33	0.002
Percentage of companies that share:			
Business plans with suppliers [F25]	75%	33%	
Cost data with suppliers [F24]	68%	11%	
Profit plans with suppliers [F26]	44%	11%	
Financial data with suppliers [F27]	19%	11%	
Characteristics of the design process:			
We design flexibility in our products [F15]	4.00	2.56	0.002
We think in terms of product platforms [F19]	4.25	3.00	0.017
Manufacturability is considered during design	5.00	4.00	0.003
[F20]	3.50	4.56	0.017
Design changes in development stage [F21]	1.50	3.90	0.000
Design changes close to production [F22]			
The transfer of learning from one NPD project to another:			
We apply the lessons learned from past projects	4.50	2.56	<0.001
[L26]	4.44	2.86	<0.001
We apply learning from post-launch meetings	1.62	3.22	<0.001
[L27]			
We repeat mistakes from one project to the other [L31]			

[a] The values given are mean scores based on a 5-point scale 1 = strongly agree and 5 = strongly disagree
[b] Significant level of difference

Table 3.14 Summary of results from Gupta & Souder (1998)

These results are consistent with literature; short-cycle-time companies work closely with their customers they use flexible highly skilled teams, and they involve their suppliers in their NPD activities. Short-cycle-time companies also design flexible products and limit design changes during the process. The most significant differences were found in the ability of short-cycle-time companies to learn from project to project.

3.2.6 R&D/Market Integration: Souder, Sherman & Davis-Cooper, 1998

Souder et al. (1998) explored the impact that R&D/marketing and R&D/customer integration had on NPD effectiveness under conditions of high and low market uncertainty. They examined 101 NPD projects in 48 firms. Twenty-five US firms participated and reported on 30 successful and 30 unsuccessful projects. Twenty-three UK firms reported on 21 successful and 20 unsuccessful projects. All firms in this study were large with more than 2,000 employees and more that US$250 million in sales. This study did not differentiate between the two countries but claimed that using data from two countries avoided results that could be attributed to national uniqueness.

NPD effectiveness was measured using the seven indicators listed below:

1. NPD Cycle time.
2. Prototype development proficiency.
3. Design change frequency.
4. R&D technical effectiveness.
5. R&D commercialisation effectiveness.
6. Product launch proficiency.
7. Market forecast accuracy.

Technical and market uncertainty were also considered as moderating factors. Market uncertainty was measured in the way described above (Souder & Song, 1997). Technical uncertainty was measured using the same scale as was used by Yap & Souder (1994).

This analysis of the INTERPROD data found that R&D/marketing and R&D/customer integration (for simplicity these will simply be referred to as integration during the short discussion below) had positive effects on NPD effectiveness, and also that these effects were moderated by technical and market uncertainty as shown in Table 3.15. The results suggest that under conditions of high technical uncertainty integration has an impact on prototype development proficiency and under conditions of high market uncertainty integration impacts the proficiency of product launch and market forecast accuracy.

	Technical Uncertain Conditions	Market Uncertain Conditions
Integration influences:	Prototype development proficiency [B52]	Product launch proficiency [B55-B57] Market forecast accuracy [C50]
Integration moderates:	Design change frequency [F21]	Prototype development proficiency [B52] Design change frequency [F21]

Table 3.15 Summary results of Souder et al. (1998)

However, they did not find that technical and marketing uncertainty moderated the effects of integration to the extent expected. This was explained by the fact that all the projects studied were totally new to the firm as opposed to product modifications or extensions and so could be classified as having high levels of uncertainty. Souder et al. (1998) conclude that high degrees of integration are necessary for success on high uncertainty NPD projects.

3.2.7 'Scrimping' in NPD: Souder, Song & Kawamura, 1998

Souder and Song (1998) use the INTERPROD data to examine the effects of 'scrimping' on new product success in the U.S. and Japan. Scrimping is defined as being less than thorough in certain areas of the NPD process, it economises the use of resources in NPD by 'getting by on a little'. This paper looks at data from 120 projects in 17 successful Japanese firms and 20 successful U.S. firms. Sixty projects from each country were examined; in each case the sample was evenly split between successful and failed new products. Three areas of 'scrimping' were 1) limited thoroughness of stage performance, 2) limited project management skills and 3) limited product success conditions.

In each case projects were examined that were thorough in each of the areas defined above and projects were examined where there were limitations in how these three areas were managed. The percentage of projects that were successful despite 'scrimping' was measured. The results of this analysis are summarised in Table 3.16.

The conclusion of this paper was that while the success rate of the Japanese firms was higher under 'ideal' conditions, the US firms were more likely to succeed when the process was not managed in a thorough manner. In other words U.S. firms can afford to 'cut corners' and have a higher chance of success that Japanese firms who operate in a similar fashion. Some of the differences found were attributed to the fact that culturally Japanese managers would be less likely to scrimp as they take a more disciplined approach to managing NPD while the US manager are more 'gung-ho' or entrepreneurial about how they manage NPD.

Area of examination	**U.S. Success Rate**	**Japanese Success Rate**
Thoroughness of Stage Performance:	70%	83%
All Stage Thorough [B50-57]	35%	14%
Development Scrimped	40%	21%
Manufacturing Scrimped	40%	18%
Marketing Scrimped		
Project Management Skills:		
All Skills High [B28-33]	65%	75%
Low Technical Skills	38%	16%
Low Marketing Skills	22%	21%
Low H.R. Skills	41%	18%
Low Leadership Skills	19%	21%
Product Success Conditions:		
All Conditions Favourable	74%	82%
Technology Ill-defined [E5]	37%	14%
User Poorly Understood [E6]	8%	8%
Low Marketing Experience [E10]	40%	14%

[a] Statistically Significant at >95% confidence level, binomial test

Table 3.16 Summary of Souder & Song (1998)

3.2.8 NPD Climate, Expertise and Management: Souder & Song, 1998

In this paper Souder & Song continue to explore the differences that they have identified between the management of NPD in the U.S. and Japan. They adopt a more thorough approach to examining the differences between the two countries. Based on their review of the literature they construct a model that examines the impact of three general dimensions, 1) NPD climate, 2) expertise and 3) management functions, on the degree of commercial success. As in previous publications, the moderating effect of market uncertainty is also considered. A guiding model, see Figure 3.3, is proposed.

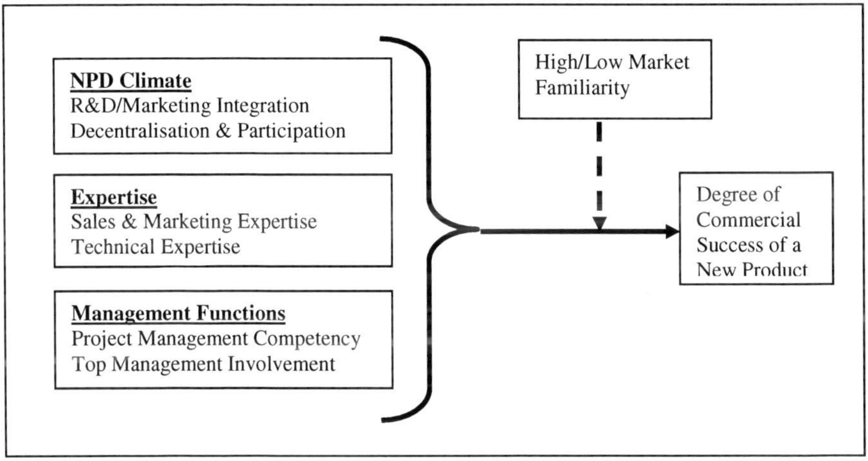

Figure 3.3 Guiding Model, Souder & Song, 1998

The sample used in this study consisted of 15 US and 15 Japanese firms, each firm reported on 60 NPD projects, 2 successes (1 for a familiar market and 1 for an unfamiliar market) and 2 failures (1 for a familiar market and 1 for an unfamiliar market). The firms were all classified as high-technology firms, operating within SIC industries 28, 35 – 38. Market certainty (or familiarity) and the degree of commercial success were measured using the same method employed in previous studies (see Yap & Souder 1994).

The three general dimensions that are explored in this study are each broken down into two multi-item measures as shown in Figure 3.3. These multi-item measures are composed of several INTERPOD questions. As in previous studies the multi-item measures were computed as the means of their component item scores. They were then correlated with the level of success of each project. Partial correlations on various combinations of the six multi-item measures were also computed. For simplicity, Table 3.17 only shows the results of the Kendall correlation between the multi-item measures and the level of new product success.

The results show a general trend of higher levels of correlation between the measures used and new product success when moving from high to low market familiarity, and also when moving from US to Japanese firms. This is consistent with the expectation that situations of low market familiarity are more demanding; in other words, it is more important to manage projects properly (i.e. high levels of competence and skills, top management involvement etc.) when they are being launched into unfamiliar markets. The correlation values were also higher for the Japanese firms; the authors explain this by suggesting that the Japanese NPD system is more firmly defined than that in the US. As suggested in their previous paper (Souder, Song & Kawamura, 1998) any variation in the system (e.g. limited skills or resources) is more likely to lead to product failure in the Japanese firms than it is in the US firms.

The authors use the information presented in Table 3.17 and the partial correlation coefficients (not presented here) to develop four precedence diagrams (one for high and low familiarity markets in each country) showing the influence of the six variables on commercial success. The key findings from these diagrams are summarised below.

- The only two measures that influence product success in US firms operating under conditions of high market familiarity are project manager competency and sales and marketing expertise.
- Under conditions of low market familiarity a different model emerges where many more measures impact on new product success. The level of R&D/marketing integration and decentralisation are also of concern and the level of decentralisation moderates the impact of project manager competency.
- Technical expertise is not found to have an impact on new product success in US firms.

Measures	Kendall Correlation Coefficients U.S. High Fam.[c]	Low Fam.[d]	Japan High Fam.[c]	Low Fam.[d]
R&D Marketing Integration: • Contact frequency between R&D and Marketing. [B18] • Information Exchanged between R&D and Marketing [B19] • Degree of harmony between R&D and Marketing [B20] • Degree of participation by Marketing [B21] • Degree of participation by R&D [B22]	.18	.38[a]	.51[a]	.77[a]
Decentralisation: • Decentralisation of project level decisions [B14] • Project personnel participation in decisions [B15] • Degree of freedom to initiate action [B16]	.23	.59[a]	.61[a]	.59[a]
Sales and Marketing Expertise: • Level of Market Research expertise [C48] • Level of Sales expertise [C49] • Level of Marketing expertise [C49]	.43[a]	.53[a]	.45[a]	.58[a]
Technical Expertise: • Level of R&D expertise [D10] • Level of Engineering expertise [D11] • Level of Manufacturing expertise [D12]	.07	.21	.40[a]	.59[a]
Project Manager Competency: • Technical Competency [B28] • Marketing Competency [B29] • Strategic Competency [B30] • Motivational Competency [B31]	.38[a]	.49[a]	.53[a]	.59[a]
Top Management Involvement: • Level of Top Management support [B23] • Top Management involvement at Project Start [B24] • Top Management involvement throughout Project [E8]	.19	.32[b]	.59[a]	.71[a]

The statistical significance for each correlation coefficient is given as: [a] $p \leq .01$, [b] $p \leq .05$, all two tailed

[c] High Familiarity Market conditions, N = 31 Japanese products and N = 30 U.S. products

[d] Low Familiarity Market conditions, N = 29 Japanese products and N = 30 U.S. products

Table 3.17 Correlation Results, Souder & Song, 1998

- The models developed for Japanese products are quite similar for both market conditions. In both cases the models are complex and success is influenced by all six measures tested.

In discussing these results Souder and Song examine the cultural differences between Japan and the US. Japan they say is characterised by institutionalised behaviour and a long-term view, while the U.S. is characterised by flexibility and a short-range orientation. Both countries have developed NPD systems to suit their cultures and it would be a mistake to attempt to copy from one country to the other.

3.2.9 UK vs. US Success Factors: Balbontin, Yazdani, Cooper & Souder, 1999

Balbontin et al. (1999) examined the INTERPROD data from 49 UK and 38 US firms operating in four industrial sectors; as in previous studies firms were asked to report on successful and failed new products. A summary of the sample of projects used in this study is shown in Table 3.18. The method used to determine project success is not detailed in this paper; however it is likely that the scoring system defined in Table 3.4 was used and that projects scoring 3 or higher were deemed to be successful.

Sector	UK Success	UK Failure	US Success	US Failures
Computers	9	3	23	20
Electronics	17	11	21	18
Chemicals	15	10	10	10
Transportation	17	17	5	2

Table 3.18 Sample analysed by Balbontin et al. (1999)

Data was analysed within five categories: Organisational, Design, Marketing, Product, and Technology. Individual items (INTERPROD questions) were analysed for their impact on project success within each of these categories – the multi-item measures used in previous publications were not developed. Market certainty/uncertainty was not considered. The responses to INTERPROD questions were tested for statistical significance (t-tests or correlation coefficients) and were categorised into three groups:

- High (H) – confidence level greater than 99.9%
- Medium (M) – confidence level between 99.9% and 99%
- Low (L) – confidence level between 99% and 95%
- Values below 95% were considered to be not significant.

Table 3.19 summarises the items that correlated at levels M or H in either country. Some of the findings that are highlighted from this study are:

- Marketing skills were correlated with success in both the US and UK.
- Project management skills were found to correlate with success in the US but not in the UK.
- In addition to the marketing skills that were common to both countries a high level of information flow between the commercial and technical entities was also correlated with success in the US and the UK.
- A close fit between existing markets and product lines and the new product was correlated with success in the UK but not the US, this is interpreted as indicating that US firms adapt themselves better to new markets than do their British counterparts.

Category[a]	**Significance for US firms**	**Significance for UK firms**
Organisational:		
Participation in decision-making of project	H	
personnel [B15]	H	H
Information flow between technical and	H	
commercial staff [B19]	H	M
Exposure of R&D personnel to business [F9]	H	
Marketing skills of project manager [B29]	H	
Management skills of project manager [B30]	H	
Authority of project manager [B33]		M
Motivational skills of project manager [B32]	M	
Technical skills of project manager [B28]	M	
Project authority concentrated in few people	M	
[B14]	M	
Partnership with component suppliers [F23]		M
Participation in problem definition by		
commercial staff [B21]		
Level of contact between technical and		
commercial staff [B18]		
Management support for innovators [B23]		
Design:		
Short cycle times in NPD process [F29 – F33]	H	
High proficiency of activities in design phase	H	

[B52]		M
High proficiency of activities in prototype		M
testing [B53]		M
High proficiency of activities in market	H	
development [B54]	M	
High proficiency of activities in marketing	M	
start-up [B56]	M	
Low number of late changes [F22]		H
Prototypes tested by users [F11]		M
Involvement of supplier staff on NPD team		
[F12]		
Designing in flexibility [F15]		
Clear project objectives [F51]		
User involvement in product definition [F6]		
Marketing:		
Adequate market research skills [C48]	H	H
Adequate sales and marketing skills [C49]	H	H
Accurate forecast of market demand [C50]	H	H
Accurate prediction of customer requirements	H	H
[C51]	H	H
Close fit between marketing skills and project		H
needs [C54]		H
Good relationship with customers [C11]		H
Market with a quick growth [C14]		M
Customers eagerness for new product [C13]		
Potential customers with great need for new		
product [C8]		
Product:		
Close fit with existing market [C52]		M
Close fit with existing product line [C53]		M
Cost effectiveness as a selling point [C42]		M
Technology:		
Good technical commercialisation [A95]	H	
Adequate R&D skills [D10]	M	
Close fit between R&D skills and project	M	
needs [D13]	M	M
Good technology exploitation [A96]	M	
Good marketing of profitable technologies		M
[A98]		M
Good development of profitable technologies		
[A97]		
Technology leadership [A101]		

[a] The items tested here are a mixture of project and firm level measures.

Table 3.19 Summary of Results, Balbontin et al. (1999)

Balbontin et al. (1999) also found that although the overall success rate of new products in the UK and the US were similar, US firms launched twice as many products as those in the UK. In summary, they point out that most of the factors related to commercial success were found to be controllable within the company

3.2.10 Market Familiarity and NPD Success: Souder & Jenssen, 1999

In this study Souder & Jenssen (1999) examine the differences between the management of NPD in US and Scandinavian (Swedish and Danish) telecommunications firms. This study also looks at the impact of market familiarity. Their sample consists of 20 US and 20 Scandinavian SBUs. In common with previous studies, each SBU was asked to report on two failed and two successful new products. The final sample consists of 70 products from each country; these 70 products are equally split between familiar and unfamiliar markets in each case. The firms were also matched in terms of annual revenue, number of employees and R&D personnel.

Market familiarity in this study was measured using 2 items as opposed to the 3 used in some previous INTERPROD publications. These items were, familiarity with the market [C5], and familiarity with customers needs [C6]. Products with a 'Yes' to both items were labelled as familiar while those with two 'No's' were labelled unfamiliar. Products that drew mixed responses to these two questions were not considered in this study. Souder and Jenssen developed a guiding model, see Figure 3.3, that proposed a particular relationship between 13 variables and new product success.

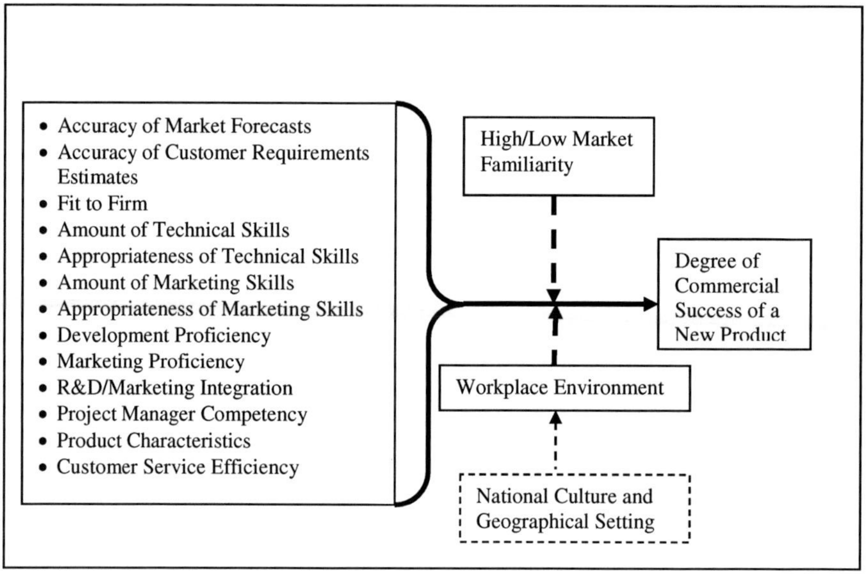

Figure 3.4 Guiding Model, Souder & Jenssen, 1999

Souder and Jenssen (1999) found that only 7 of the 13 variables outlined in Figure 3.4 impacted on new product success within the firms studied. Their results for these 7 variables are detailed in Table 3.20.

	Correlation Coefficients US Products		Scandinavian Products	
	Familiar Market Products	Unfamiliar Market Products	Familiar Market Products	Unfamiliar Market Products
Market Proficiency: Proficiency of Market Development [B54] Proficiency of Market Launch [B56] Proficiency of Market Research Level of Sales and Marketing skills [C49] Fit of Sales and Marketing skills [C54] Accuracy of Market Forecast [C50] Accuracy of customer req. [C51]	.67[c]	.72[c]	.58[c]	.58[c]
Technical Skill Adequacy: Amount of R&D skills [D10] Amount of engineering skills [D11] Amount of manufacturing skills [D12] Fit of R&D skills [D13] Fit of engineering skills [D14] Fit of manufacturing skills [D15]	NS	.39[a]	NS	.45[b]
R&D/Marketing Integration: Frequency of contact [B18] Amount of information flow [B19] Marketing participation [B21]	NS	.69[c]	NS	NS

Project Management Competency: General management skills [B30] Participation skills [B31] Motivational skills [B32] Degree of authority [B33]	NS	.49[b]	NS	NS
Product Characteristics: Quality [C36] Performance superiority [C38] Technical superiority [C39] Reliability [C45]	NS	NS	NS	.45[b]
Development Proficiency: Proficiency of exploratory research [B50] Proficiency of concept development [B51] Proficiency of prototype develop. [B52] Proficiency of prototype testing [B53]	.48[b]	.46[b]	.42[a]	.41[a]
Customer Service Efficiency: Adequacy of customer support [C43] Efficiency of technical services [B57]	.41[a]	.51[b]	.38[a]	.56[c]

Note: N = 35 for each group pf products; coefficients given are Spearman correlation coefficients.
[a] $p \leq .05$ (two-tailed); [b] $p \leq .01$ (two-tailed); [c] $p \leq .001$ (two-tailed);

Table 3.20 Summary of Results, Souder & Jenssen, 1999

The results above show that in the US three variables; 1) market proficiency, 2) development proficiency and 3) customer service efficiency are correlated with product success under both familiar and unfamiliar market conditions. Three other variables; 1) technical skill adequacy, 2) R&D/marketing integration and 3) project management competency also correlate under conditions of market uncertainty. These results are consistent with previous findings, some based on INTERPROD (Yap & Souder, 1994) and others (e.g. Cooper & Kleinschmidt, 1995). In particular, it is worth noting that the general trend seen in many of the INTERPROD publications, whereby management practices are of more importance when developing new products for unfamiliar markets is also supported in this study.

One unusual finding in the above table is the lack of correlation between product characteristics and success in the US. The authors explain this by noting that in familiar markets responses to questions about product characteristics were bimodal, with two types of products being identified. These are described as 'low-frill' and 'high-frill' products. The low-frill products would typically not score highly on the product characteristics measured - see Table 3.20. The high-frill products would tend to be of a higher quality, reliability, performance and technical competency. These two types of product would be aimed at different market sectors but both were seen to have similar chances of success, thus explaining the lack of correlation.

By contrast, in unfamiliar markets US manufacturers were found to over-develop products for unfamiliar markets; in other words, they were more likely to develop high-frill products for unfamiliar markets. The results of this study suggest that this is not a particularly successful strategy.

The same three variables 1) market proficiency, 2) development proficiency and 3) customer service efficiency were found to correlate with success in both familiar and unfamiliar Scandinavian markets. Technical skills were also found to correlate with success in unfamiliar but not familiar markets, the same pattern that was seen in the US data. However, there are some differences between the US and Scandinavian results. Neither R&D/marketing integration nor project manager competency correlate with success under any market conditions in the Scandinavian firms; additionally, product characteristics were found to relate with success under unfamiliar market conditions.

The authors discuss in considerable detail the differences between the workplace in the US and Scandinavia. The findings of this paper and previous publications (Hofstede, 1991) suggest that the national culture in Scandinavia is 'feminine' (stressing humanisation of work, mutual assistance and concern for welfare) while that in the US is more 'masculine' (stressing assertiveness, results and competition). This results in a workplace environment in Scandinavian firms that is characterised by enduring relationships and mutual dependency. In such an environment there is no need to actively encourage R&D/marketing integration. Additionally, the need for an individual with high management and technical skills is less important in Scandinavia as the NPD team is likely to be internally coordinated and controlled. This theory is supported by the responses to several other INTERPROD questions which show that in Scandinavian teams 'anyone can lead', and 'persons typically assume whatever roles are needed'.

One of the most important conclusions of this paper is that the results suggest that global NPD theories may be inappropriate and that many basic NPD principles may need to be adjusted for cultural differences.

3.2.11 NPD Organisational Characteristics: Lee, Lee & Souder, 2000

This study compares the different organisational settings for NPD in the US and Korea. Three questions are posed:

1. Are there differences in the organisational characteristics of the NPD process in Korea and the US?
2. Are there differences in the impact of organisational characteristics on NPD success in Korea and the US?
3. If differences exist, what are the reasons for them and what are the management implications?

A model is proposed which links organisational factors to NPD success. All the variables tested in this paper were collected using the INT-B questionnaire. Figure 3.5 shows the proposed model and identifies the INTERPROD questions used to collect each data item. Commercial performance was measured in the same way as in earlier INTERPROD papers and this paper did not consider market familiarity as a variable.

The authors examined the organisational structures used for NPD projects in Korea and the US and found that the use of a NPD department (61.5% of projects) and a NPD manager (34.6% of projects) were most popular in Korea, while a NPD manager (44.2% of projects) and top management control (30.2% of projects) were the most popular methods in the US.

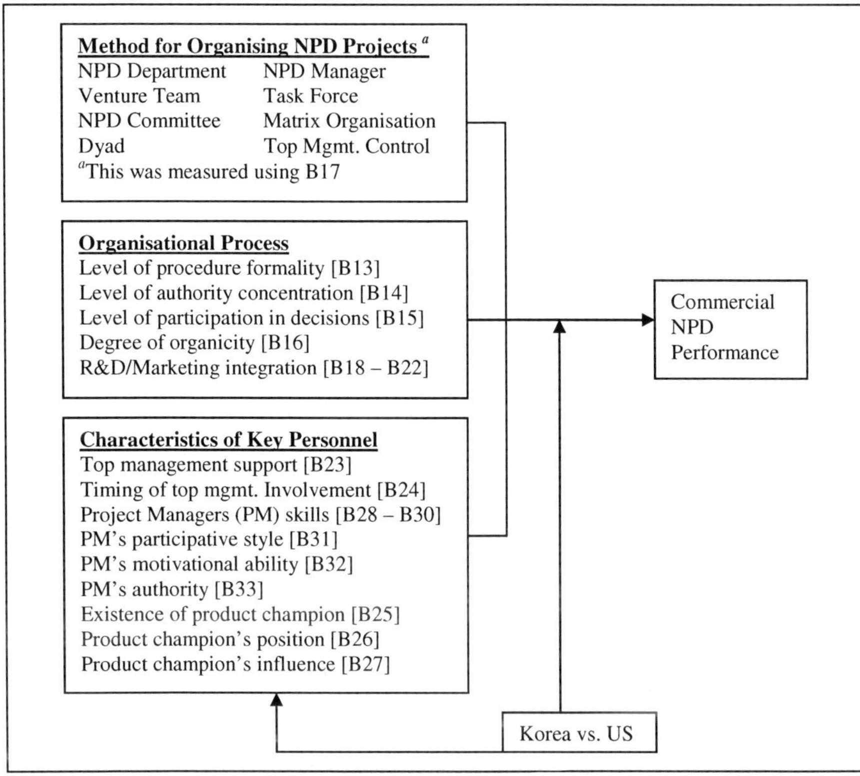

Figure 3.5 Conceptual Framework, Lee, Lee & Souder, 2000

The mean values of the organisational process and the characteristics of key persons items (see Figure 3.5) in the US and Korea were compared and significant differences (*t*-test, $P < 0.1$) were noted. The degree of R&D/Marketing integration was lower in Korea than in the US, as was the level of authority concentration.

Top management in Korea were found to get involved at an earlier stage in NPD projects. The level of project manager's skills and the degree of project manager authority were both lower in Korea. The authors attributed many of these differences to the high level of collectivism in Korea.

The next stage of analysis in this paper was to correlate the factors described in Figure 3.5 with NPD success. The authors carried out a subgroup analysis on the methods used for organising NPD projects. The samples for each country were spilt into two based on whether or not they had used a particular method of organisation. The average level of success of each group was then compared. The use of venture teams was found to lower the level of performance in Korea but to increase it in the US. The use of a matrix organisation was found to increase success in Korea.

The mean values of the organisational process and the characteristics of key persons items were correlated with success for both countries. Significant differences were found between the two countries in three items: 1) level of authority concentration, 2) degree of organicity, and 3) project manager's participative style. Concentration of authority was found to facilitate success in the US but not in Korea. A high level of organicity and a project manager with a participative style were found to be more closely linked with success in Korea than the US. It is also interesting to note that the characteristics of the project manager correlated with success in both countries.

The overall conclusion of this paper is that country-specific culture does have an impact on how NPD should be managed. The authors propose a positional model which details the direct impact of culture on organisational characteristics, institutional management systems and NPD performance, but also shows the indirect link that culture has through the impact of institutional management systems.

3.2.12 Market Orientation: Wren, Souder & Berkowitz, 2000

This paper reports on the impact of market orientation in six of the countries involved in INTERPROD. Market orientation is defined as 'the set of cross-functional processes and activities directed at creating and satisfying customers through continuous needs assessment'. The countries involved in this study are the US, New Zealand, Korea, Scandinavia (Sweden and Norway), and Belgium. Four hypotheses are put forward:

Hypothesis 1. The degree of new product success will be positively related to:

a. The level of marketing skills adequacy
b. The level of technical skills adequacy

Hypothesis 2. The degree of new product success will be positively related to:

a. The level of market intelligence
b. The degree of customer orientation

Hypothesis 3. The level of market intelligence will be positively related to:

a. The project manager's marketing management and motivational skills
b. Top management support for the project
c. The level of cross-functional integration on the project

Hypothesis 4. The degree of customer orientation will be positively related to:

a. The project managers marketing management and motivational skills
b. Top management support for the project
c. The level of cross-functional integration on the project

Theses hypotheses were tested using the INTERPROD items shown in Table 3.21.

	U.S. (N = 111)	New Zealand (N = 74)	Korea (N = 52)	Scandinavia (N = 72)	Belgium (N = 66)
Dependent Variable = Success					
Market Intelligence:	.57[a]	.42[b]	.31[b]	.28[b]	.34[a]
Forecast of market demand was accurate [C50]					
User requirements well understood [E6]					
Customer Orientation:	.21[b]	.29[b]	.24	.32[b]	.38[a]
Cust. problems with previous products considered [C23]					
Customer's needs were well defined [C6]					
Marketing Skills Adequacy:	-.19[c]	-.07	.06	-.16	-.01
Market research skills were ideal for the project [C48]					
Sales force skills were ideal for the project [C49]					
Technical Skills Adequacy:	.03	-.01	.29[b]	.19[c]	.18[b]
Manufacturing skills were ideal for the project [D12]					
Engineering skills were ideal for the project [D11]					
Dependent Variable = Market Intelligence					
Project Management Skills:	.21[b]	.02	.34[b]	.22[c]	-.05
Project Manager could motivate the team [B32]					
Project Manager has a participative style [B31]					
Top Management Support:	.22[b]	.28[b]	-.03	.12	.12
Top management was supportive of the project [B23]					
Top management involvement in the project was high [E8]					
Integration:	.26[b]	.19[c]	.29[b]	.30[b]	.59[a]
Level of contact between tech. and comm. entities [B18]					
Information flow between tech. and comm. entities [B19]					

Dependent Variable = Customer Orientation					
Project Management Skills (measured as above)	.29[a]	.14	.01	.28[b]	-.01
Top Management Support (measured as above)	.13	.22[c]	.13	.09	.31[b]
Integration (measured as above)	.29[a]	.23[b]	.40[a]	.19[c]	.22[c]

[a] $p < .01$ (two-tailed); [b] $p < .05$ (two-tailed); [c] $p < .10$ (two-tailed);

Table 3.21 Summary Results, Wren, Souder & Berkowitz, 2000

The results in Table 3.21 show that Hypothesis 1b is weakly supported, the only countries in which technical skills adequacy is important are Korea, Scandinavia and Belgium and even here the correlations are not strong. Marketing skills adequacy does not correlate with success in any of the countries studied – Hypothesis 1a is not supported. Hypothesis 2 is supported by the results in Table 3.21, customer intelligence and market intelligence correlate with success in all countries with the exception of customer intelligence in Korea.

Hypothesis 3 is weakly supported in all countries (two of the three items tested show significant correlation coefficients) with the exception of Belgium. Hypothesis 4 is also supported in all countries with the exception of Korea.

The key findings that the authors drew from these results are listed below.

- Marketing intelligence is a critical skill for all firms regardless of country of origin.
- Customer orientation is also a significant factor for most countries, with the exception of Korea. The authors speculate that the 'collective' culture of Korea makes it different from the other countries involved in the study and that individualist countries are more likely to support customer orientation.
- Technical skills are required to a greater extent in countries that have a disposition towards exporting.
- Integration is a critical component of both market intelligence and customer orientation.
- In the US and New Zealand, project manager skills and top management support are positive predictors of market intelligence and customer orientation.

The authors conclude by pointing out that although there are many consistencies across countries; there are also differences and these have implications for managers. Three primary implications are proposed.

1. The integration of teams is critical to gaining market intelligence and a customer orientation and thus producing successful new products.
2. Culture is an important factor in determining the level of managerial support and skill necessary to develop a customer orientation. Hofstede's individualism index provides a rationale for why some countries are more customer-oriented than others.
3. Technical skills are important but not critical, some countries, for example the US, have a culture of hiring in skills when needed, while others are less likely to do this.

3.2.13 Cross-Functional Integration: Sherman, Souder & Jenssen, 2000

This study focuses on cross-functional integrations and the impact that it has on product development cycle time. INTERPROD results from 65 US and Scandinavian firms are compared, the sample includes firms from a range of industries including telecommunications, computers, instruments, chemicals, biotechnology and software.

Seven variables were defined, each calculated as the mean for a number of INTERPROD items. Table 3.22 summarises the composition of the variables used, their means, standard deviations and correlation coefficients.

Table 3.22 shows that with the exception of strategic partnership integration, all the integration measures were found to be positively related to cycle time. The authors carried out a series of stepwise multiple regressions to test the relative contributions of the various integration variables. The results of these tests led the authors to conclude that integration of knowledge from past projects has the greatest impact on cycle times, followed by R&D/marketing integration and R&D/customer integration. A relationship between integration of knowledge from past projects and R&D/customer integration was found but the authors concluded that this was not causal. They argued that firms who emphasized learning were also likely to emphasise R&D customer integration. Further tests, one using dummy variables for product categories and a second using a dummy variable for region, were conducted and the results showed that there were no significant differences between products from different industrial sectors or regions. In other words the results were found to be generalisable across both sector and region.

Variable	Mean	SD	Correlation[a] 1	2	3	4	5	6	7
1. Integration of Knowledge from Past Projects	3.50	.88	-						
Lessons learned from past projects are shared [L15]									
Post-launch meetings are conducted [L16]									
Incisive decisions occur from post-launch meetings [L17]									
We apply lessons from past projects [L26]									
We apply information from post-launch meetings [L27]									
2. R&D/Customer Integration	3.73	.85	**.64**	-					
We rely on users to define needs [F6]									
We often have users try out developments [F8]									
We give working prototypes to users [F11]									
We review customers reactions [L11]									
3. R&D/Marketing Integration	3.56	.59	.15	.24	-				
Contact between R&D and marketing [B18]									
Information flow between R&D and marketing [B19]									
Participation by marketing in problems [B21]									
Participation by R&D in problems [B22]									

4. R&D/Manufacturing Integration Manufacturing details are considered [F20] Close R&D manufacturing integration [F58] Attention to manufacturability [F40]	3.76	.75	**.39**	**.45**	.24	-			
5. R&D/Supplier Integration Component suppliers on our development teams [F12] Use competitors parts if necessary [F13] Partner with component suppliers [F23]	3.22	.99	**.52**	**.49**	.19	**.39**	-		
6. Strategic Partnership Integration Involvement of universities [A36] Involvement of Research institutes [A37] Involvement of competitors [A38] Involvement of other companies [A39]	2.49	.69	-.10	.00	.11	-.01	-.06	-	
7. Cycle Time Our cycle times are much shorter than our competitors [F29 – F33]	3.17	.89	**.63**	**.50**	**.33**	**.42**	**.45**	-.04	-

[a] Pearson product moment correlations greater than .24 at n = 65, significant at the .05 level (two-tailed)

Table 3.22 Summary of Results, Sherman, Souder & Jenssen, 2000

3.3 Summary of INTERPROD Publications

Table 3.23 summarises the 13 INTEPROD publications, showing the focus of the study, for each the manner in which the data were split, the factors analysed and the key findings of each publication. Some of the common themes that have emerged are:

1. The NPD strategies that work in large firms do not always work in small firms. High-tech small firms are a specific category of

firms and it should not be assumed that what works in large technology-based firms will work in small technology-based firms. Studies that looked at both firm size and location found that size had a bigger impact on determining success strategies than did location.

2. Firm location (country) has an impact on the way in which NPD is managed and also on the strategies that are most effective. Many similarities were found between the different countries studied but there were also significant differences.
3. Products developed for markets that are not well-defined or for situations where customer needs are not well-known and cannot easily be translated into technical specifications, are more difficult to manage. They require higher levels of skill and proficiency.
4. Marketing skills, proficiency were linked with success in most countries and under most conditions. The link between technical skills and proficiency and success was less clearly defined – in some studies a link was found but not in others.
5. The impact of organisational factors (e.g. top management participation, concentration of authority etc) and project management characteristics on product success varied from country to country, and also with the level of market certainty.
6. The link between R&D/marketing integration and new product success was supported by most of the publications.
7. Learning from past projects was found to be linked with high levels of integration and also with reduced NPD cycle time.

The position of each of the thirteen papers reviewed within the overall INTERPROD framework is shown in Figure 3.6. From this it is clear that most of the publications to date have focused on project level NPD management and various influencing factors. Only two papers (Gupta & Souder, 1998; Sherman et al., 2000) examined company level NPD management. Firm location has been the influencing factor examined by most of the papers. Only three papers (Yap & Souder, 1994; Souder & Song, 1997; Souder, Buisson & Garret, 1997) examined issues relating specifically to small firms.

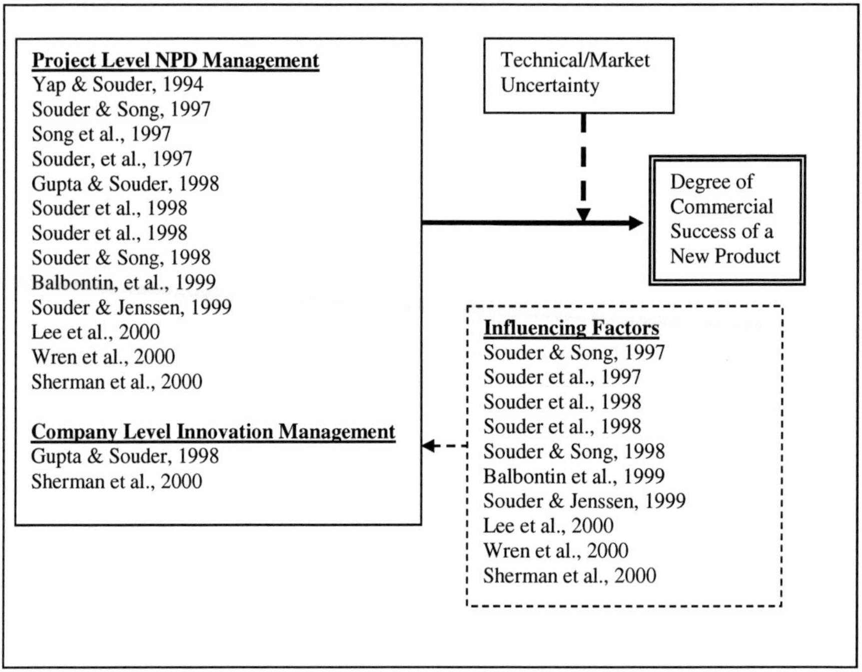

Figure 3.6 Previously published papers within the INTERPROD framework.

Publication	Key Theme	Data Split	Factors Analysed	Key Findings
1. Yap & Souder, 1994	Success and failure in small high-tech electronics firms	Market uncertainty Technical uncertainty	Technology Sources Product Characteristics Market Characteristics Entry Strategies Organisational Characteristics	Small high-tech firms are not the same as large firm and need to modify their strategy based on the level of market uncertainty.
2. Souder & Song, 1997	Product design and marketing strategies in small and large, US and Japanese, electronics firms.	Market uncertainty Firm size Firm location	Performance superiority Technical superiority Radicalness Compatibility Users eagerness to adopt Market growth potential Intensity of competition	Firm size has a greater impact on successful NPD strategies than firm location. Designing compatible products is a good strategy for all firms in certain markets but only for small firms in uncertain markets. In uncertain markets the same effective strategies apply for large US and Japanese firms, design radical superior products for eager users in high growth markets. Designing compatible products for competitive markets is always an effective strategy for small firms.
3. Song, Souder & Dyer, 1997	Skills, technical and market proficiency and product quality in large high-tech Japanese firms		Product Quality Market proficiency Technical Proficiency Process Skills Project management skills Skill/needs alignment Team skills Design sensitivity	New product success was positively correlated with product quality; this, in turn, was linked to marketing proficiency. Technical proficiency was not found to be linked to product quality or new product success. Marketing skills depended on process and project management skills and skill/needs alignment. Technical skills depended on team skills and design sensitivity but not on skill/needs alignment.

4. Souder, Buisson & Garret, 1997	Customer driven NPD in small US and New Zealand high-tech firms	Market uncertainty Firm location	Market proficiency Technical Proficiency Technical skills Marketing skills Project manager characteristics R&D/marketin g integration	New Zealand firms are more successful due to their closer relationships with their customers. Marketing proficiency is more closely linked to product performance than technical proficiency. Technical skills were not linked to product success. Project manager characteristics were more closely linked with success in the US than they were in New Zealand.
5. Gupta & Souder, 1998	Benefits of reduced cycle time in US high-tech firms	Length of life-cycle	User involvement in NPD Management of NPD teams Supplier involvement Design process Transfer of learning	Short life-cycle firms were found to be more profitable and have higher NPD success than other firms. They were found to design more flexibility into their products, work more closely with customers and suppliers and, most importantly, learn from past projects.
6. Souder, Sherman & Davis-Cooper, 1998	Uncertaint y and organisatio nal integration in large, high-tech US and UK firms	Market uncertainty Technical uncertainty	Prototype level proficiency Product launch proficiency Market forecast accuracy Design change frequency	In conditions of technical uncertainty, prototype development proficiency impacts integration; in high market uncertainty, launch proficiency and forecast accuracy impact on integration. High degrees of integration (R&D/marketing and R&D/customer) are necessary for NPD success.
7. Souder, Song & Kawamura, 1998	Controlled 'scrimping ' in US and Japanese high-tech firms	Firm location	NPD stage performance Project Management Skills Product success conditions	US firms can gain a competitive advantage by 'scrimping' on NPD projects. In other words, US firms perform better under less than ideal conditions.

8. Souder & Song, 1998	NPD climate, expertise and management functions in US and Japanese high-tech firms	Market uncertainty Firm location	R&D/marketing integration Decentralisation Sales and market expertise Technical expertise Project manager competency Top management involvement	Only sales and marketing expertise and project manager competence correlated with success under all conditions. Technical expertise only has an impact on success in Japanese firms. NPD is more demanding in unfamiliar markets, firm location impacts on successful NPD strategies.
9. Balbontin, Yazdani, Cooper & Souder, 1999	Success factors in US and UK high-tech firms	Firm location	Organisation Design Marketing Product Technology	There were many similarities between both countries; differences were in the impact of Project Management Skills which were more important in the US, and ability to adapt to new markets which was higher in the US.
10. Souder & Jenssen, 1999	NPD management practices in US and Scandinavian high-tech firms	Market uncertainty Firm location	Market proficiency Technical skills adequacy R&D/marketing integration Project management skills Product characteristics Development proficiency Customer service efficiency	Market proficiency, development proficiency and customer service efficiency correlate with success under all condition – but cultural differences were observed. Technical skills, integration and project management only had an impact in unfamiliar US markets. Technical skills and product characteristics only impacted products in unfamiliar Scandinavian markets.

11. Lee, Lee & Souder, 2000	Organisational characteristics in US and Korean high-tech firms	Firm location	Organisation for NPD projects Organisational processes Key personnel	Organisational process and project manager characteristics were linked with success in both countries. Organic structures and participative project managers were linked with success in Korean firms. High concentration of authority was linked with success in US firms. Project Management Skills were linked with success in both countries.
12. Wren, Souder & Berkowitz, 2000	Market orientation in US, New Zealand, Korea, Scandinavia and Belgium high-tech firms	Firm location	Market intelligence Customer orientation Marketing skills Technical skills Project Management Skills Top management support Integration	Market intelligence is critical for NPD success in all countries. Customer orientation is linked with success in all countries except Korea. Integration is critical for market intelligence and customer orientation. Project Management Skills impacted on market intelligence in US, Korean and Scandinavian firms.
13. Sherman, Souder & Jenssen, 2000	Integration in US and Scandinavian high-tech firms	Firm location Industrial sector	Integration with past projects R&D/customer integration R&D/marketing integration R&D/manufacturing integ. R&D/supplier integration Strategic partnership Cycle time	With the exception of strategic partnerships all variables tested correlated with NPD success. The strongest link was between integration of knowledge from past project and success. R&D/marketing integration and customer integration were found to be more important than supplier or R&D/manufacturing integration. These results applied broadly across locations and sectors.

Table 3.23 Summary of INTERPROD publications

3.4 Conclusion

One key limitation of the above analysis is that the factors examined in each study were not analysed in a consistent manner. This means that although some general observations can be made about the overall results of the 13 INTERPROD papers, direct comparisons are difficult. However, several of the general themes outlined above are of interest to the management of NPD projects in small Irish high-tech electronics firms. The next chapter proposes a model for the management of NPD in Irish electronics firms based on INTERPROD and the literature reviewed in Chapter 2. This model will attempt to explain the impact of firm size on the management of NPD and also on NPD success factors.

CHAPTER FOUR

MODEL DEVELOPMENT

4.1 Introduction

At the start of Chapter 2 a model was proposed illustrating, at a high level, the research question for this study. It is now possible to further develop this model based on the literature reviewed in Chapter 2 and the INTERPROD project described in Chapter 3. The research model that will guide the rest of this study is presented in Figure 4.1.

The variables included in this model have been divided into those that apply at a *company level,* for example the amount of resources allocated to developing new products or the ability of the firm to launch new products on time, and those that apply at a *project level* for example the characteristics of the project manager or proficiency at the NPD process. This is to maintain compatibility with the INTERPROD model. Also, while the simplified model presented at the start of Chapter 2 considered only firm size, the main focus of this thesis, as moderating NPD management and success, the model above considers the *NPD context* as a moderating parameter. Here the NPD context is taken to mean the overall context within which firms innovate. This reflects the findings of the INTERPROD project; firm location (country) and market/technical certainty all have a moderating effect on the management and success of NPD.

The variables listed in Figure 4.1 can be grouped into 3 general area as shown in Table 2.2; Strategic and Market Factors, Development Process Factors and Organisational Factors. However, at this point no other grouping of variables or relationships between variables have been identified.

The rest of this chapter will review each of the variables included in the model; they have already been described in detail in Chapters 2 and 3. The following topics will be covered:

1. NPD Context – Section 4.2.
2. Measures of NPD Success – Section 4.3.
3. Company level determinants of Success – Section 4.4.
4. Project level determinants of Success – Section 4.5.

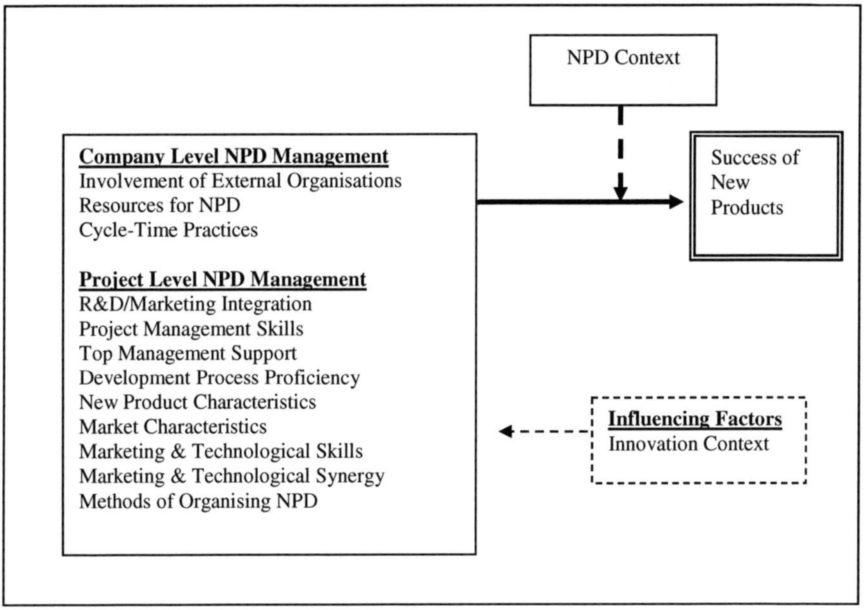

Figure 4.1 Proposed Research Model

Each topic will be discussed briefly to determine how they are expected to affect new product success and the possible effect of firm size on their management and impact. Section 4.6 contains the conclusion of this chapter.

4.2 NPD Context

This variable examines the context within which large and small firms develop new products. It addresses issues such as firm size and turnover, percentage of export and the key concerns of each group of firms. The key variable of interest in this study is firm size however, it is also interesting to look at the impact of other contextual variables on new product success. The literature reviewed in Chapter 2 would suggest that firm size should have an impact on:

- The formality of the NPD process and the level of product innovation (Walsh et al. 2002; Roper, 1996; Santarelli & Steracchini, 1990).
- The new product and market characteristics that are linked with success (Yap & Souder, 1994).
- The amount and nature of resources available for NPD (Roper et al. 1996; Hoffman et al. 1998; Love & Roper, 2001).
- The type of NPD process that is most effective (Bessant & Francis, 1997).
- The ability of firms to reduce new product development time (Walsh et al. 2002; Ali et al. 1995)
- The nature of R&D/Marketing interaction (Rothwell & Dodgson, 1991)
- The importance of top management skill or support (Hofmann et al. 1997)

4.3 Measures of New Product Success

The success of a new product can be measured in several different ways: (1) financially (profit, sales, payback period, costs), (2) market share and (3) technical performance (Montoya-Weiss & Calantone, 1994). The first two of these measures, financial and market, can be taken together as measure of the commercial success of a new product project. Many strategies for measuring NPD success have been developed by different authors, for examples see Kaplan & Norton, 1992 or McGrath & Romeri, 1994. However the way in which success is measured by academics does not always reflect the opinions of industrialists (Driva et al. 2000).

One of the difficulties in devising a measure of NPD success in a study which plans to compare NPD in small and large firms is that small firms may have very different project success criteria than large firms. A way to overcome this problem is to define success in terms of the original project objectives, in other words – 'a project will become successful only after meeting the objectives and expectations envisaged at the time of its selection' (Pillai et al. 2002). This method of measuring NPD success based on the original project objectives and strategy has been found to be acceptable both by practitioners and academics (Griffin & Page, 1996) and so will be adopted in this study.

4.4 Company Level Determinants of Success

4.4.1 Involvement of External Organisations

There is little consensus within the literature surveyed on the impact of external interaction on new product success. Some sources suggest that involving external organisations in NPD projects is linked with success, particularly in small firms (Eurostat, 2000; Rothwell & Dodgson, 1991) while others find little evidence to support this view (Freel, 2003; Love & Roper, 2001). Interactions with several different types of organisations will be measured to determine the extent of involvement of external organisations and also the impact of this involvement on new product success. The differences between small and large firms in terms of the level, impact and nature of involvement of external organisations will also be measured.

4.4.2 Resources for NPD

A lack of resources, both financial and human has been cited as a key limitation for small firms developing new products (Fitzgerald and Breathnach, 1994; Roper et al., 1996; Hoffman et al., 1998). This variable will be used to examine the differences between small and large firms in terms of the resources they utilise in developing new products and also how these resources are allocated.

4.4.3 Cycle-Time Practices

Shorter new product development times are linked with increased commercial success (Ali et al., 1995; Vesey, 1991). There is also evidence to suggest that small firms are faster at developing new products than large firms (Walsh et al., 2002; Ali et al., 1995). Many studies have attempted to identify management practices that are effective in reducing NPD cycle time but most have examined only large firms (Griffin, 1997; Zirger & Hartly 1996; Eisenhardt & Tabrizi, 1995). The cycle-time performance of small and large firms will be examined along with the effectiveness of a variety of cycle-time reduction techniques.

4.5 Project Level Determinants of Success

4.5.1 R&D/Marketing Integration

Several studies have established a link between R&D/Marketing integration and new product success (Griffin & Hauser, 1996; Song & Parry, 1997; Leenders & Wierenga, 2002). However the general perception is that in small firms R&D and marketing personnel communicate well (Berry & Taggart, 1998; Souder, 1995). The degree of integration between R&D and marketing and also the impact that this has on new product success will be measured.

4.5.2 Project Management Skills

Project manager behaviour and competence has been identified as a success factor in product development (Norrgren & Schaller, 1999; Yap et al. 1998). Earlier studies also linked project manager's position or authority with success (Wheelwright & Clark, 1992). Within this study both the skills and authority of project managers will be examined to determine the degree to which they have an impact on new product performance in small and large firms.

4.5.3 Top Management Support

Top management support is linked with new product success in many studies (Swink, 2000; Brown & Eisenhardt, 1995, Montoya-Weiss & Calantone, 1994) but the effect of top management support in small firms is not as clear (Hoffman, 1998). This variable will be measured to determine the relative importance of top management support in small and large firms.

4.5.4 Development Process Proficiency

Several studies have attempted to determine which of the various stages in the product development process are most important for new product success (Song & Montoya-Weiss, 1998; Cooper & Kleinschmidt, 1987). The relationship between proficiency at different NPD activities and success will be measured in small and large firms. Additionally the degree to which formal processes are used and linked with success will also be examined.

4.5.5 New Product Characteristics

Developing technically superior products is linked with success in large firms (Cooper, 1990) but not always in small firms (Yap & Souder, 1994). A range of new product characteristics is examined to determine which characteristics are linked with success in small and large firms. Several studies have suggested that small firms develop mainly incremental rather than radically new products (Walsh et al., 2002; Santarelli & Sterlacchini; 1990). The degree to which small and large firms are involved in radical and incremental, and the impact of this on success will be explored.

4.5.6 Market Characteristics

There are conflicting opinions about the impact of market maturity and competitiveness on new product success. Some studies have suggested that small firms have an advantage in newer markets (Acs & Audretsch, 1988) while others have claimed the opposite (McGrath, 1995). It is generally assumed that mature markets offer few new product opportunities (Porter, 1980) but more recent studies have found that small firms are successful in such markets (Yap & Souder, 1994). The impact of these market characteristics on new product success in small and large firms will be examined.

4.5.7 Marketing and Technological Skills

As with market and technological synergy, there is general consensus that market and technological skills are linked with new product success (Wren et al., 2000; Zirger & Maidique, 1990). This study will examine whether the skills linked with success in small firms are the same as those linked with success in large firms.

4.5.8 Marketing and Technological Synergy

New product that are developed in a manner that exploits a firm's existing technological and marketing skills are found by many researchers to be more successful than those that do not (Song & Parry, 1999; Balbontin et al., 1999). The extent to which marketing and technological synergy is linked with success in small firms will be examined in this study.

4.5.9 Methods of Organising NPD

A variety of organisational structures are used for NPD projects, with some being more closely linked with success than others (Lee et al. 2000). This study will examine the methods used by small Irish electronics firms in managing NPD projects and identify those that are most closely linked with new product success.

4.6 Conclusion

The model that will guide this research was presented at the start of this chapter. The model is broadly based on the INTRPROD project but has been modified to include firm size as a variable that affects the context within which firms innovate and as such has an impact on the manner in which NPD is managed and also on the determinants of new product success.

The management variables that populate this model were identified in the literature review in Chapter 2 and have been briefly considered again in this chapter to determine the possible impact of firm size.

The next chapter will describe the research method that will be used to test the ability of this model to describe and explain the management of product development in small Irish electronics firms.

CHAPTER FIVE

RESEARCH DESIGN

5.1 Introduction

This chapter describes the plan for data collection and analysis to be used in this study. The research model presented in Chapter 4 is exploratory; the aim of this study is not to prove any specific hypothesis but to examine how small firms manage NPD within an Irish setting. To this end a multi-method research strategy has been selected incorporating a survey and a set of case studies. This chapter begins, in Section 5.2, with an introduction to the research philosophy underpinning this investigation. Section 5.3 presents the research methods to be used and the rationale behind the selection of these methods. Sections 5.4 and 5.5 describe in detail both methods, questionnaire and case study, in terms of the data collection and analysis used for each. An evaluation of the research methods selected in terms of validity and reliability is presented in Section 5.6 and finally the conclusion of the chapter is in Section 5.7.

5.1.1 Research Objectives

The objective of this research is to determine the influence of firm size on new product development in Irish electronics firms. The primary research question to be answered is:

How do small Irish electronics firms manage the development of new products, and are the success factors that apply to small Irish firms different from those that apply to large Irish firms?

This research question has been extended through the development of the research model presented in Chapter 4. Using the research model the objectives of this study can be operationalised facilitating the development of a research plan.

The model, presented in Figure 4.1, proposes a link between twelve NPD management variables and new product success. It also suggests that the way in which NPD is managed and the impact of the twelve variables on new product success are dependent on firm size and location. Based on this model the research question will be answered by achieving the following objectives.

1. To determine the differences between the management of the twelve NPD management variables in small and large Irish electronics firms.
2. To determine, using these twelve NPD management variables, which management practices are linked with success in small and large Irish electronics firms.
3. To explore the extent to which the management of NPD in Ireland is different from that in other countries.
4. To explore the extent to which the differences between small and large firms in Ireland are also seen between small and large firms in other countries.

The research model is being used to guide and structure the research it should not be viewed as a hypothesis or theory to be tested. This study is exploratory and the aim is to develop a theory about the management of NPD in small firms.

5.2 Research Philosophy

In social science two fundamentally different approaches to the research process exist. The first, known as *positivism*, relies on an objective view of knowledge. Data or 'facts' are collected in a value-free manner and analysed statistically to develop universal causal theories and laws. The second approach appears under many different labels; naturalism, relativism, phenomenology, constructivism – for the purpose of this discussion it will be referred to as *relativism*. This approach, in its extreme form, maintains that reality does not exist outside human consciousness. A more practical approach to relativism assumes that researchers are subjective and influence the data that they collect, also that research generates hypotheses rather than immutable facts (Robson, 2002). A brief description of each of these research approaches follows.

5.2.1 Positivism

The positivist approach relies on the research philosophy of natural or physical scientists. Knowledge is treated objectively, in other words human practices and behaviours can be defined by quantifiable variables. These variables can be studied independently and theories and laws can be developed to predict the future. The researcher is assumed to be totally objective neither affecting nor being affected by the research subject (Saunders et al., 2000). This approach reduces human actions to automatic responses to external stimuli (Tiernan, 1995). Positivism relies on highly structured research methods that are easily replicated and produces data that are suitable for statistical analysis (Gill & Johnson, 1997). The research approach generally taken by positivists is to develop a theory and attempt to prove it using *quantitative* methods.

5.2.2 Relativism

Relativists reject the positivist approach on the grounds that the social world of, for example, management is too complex to be defined by a set of variables and rigid laws. Research should usefully generate models that can be evaluated in terms of their ability to describe as well as to predict. Relativists view reality through the eyes of their research subjects, and consider each research setting to be unique. The ability to generalise is not of critical importance to relativists (Saunders et al., 2000). Relativists tend to use unstructured research methods that can capture multiple views and aspects of a situation. The research approach taken by relativists is to generate, rather than prove, theory or hypotheses based on data using *qualitative* methods.

The descriptions of positivism and relativism above present two different uses of data. From a positivist perspective data are collected to prove a preconceived theory while from a relativist perspective data are collected to develop theory. These two approaches are represented as the two sides of Kolb's experiential learning cycle (Gill & Johnson, 1997) as shown in Figure 5.1.

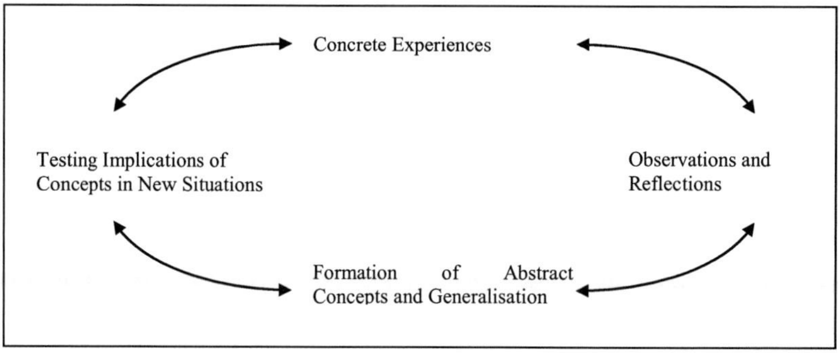

Figure 5.1 Kolb's Experiental Learning Cycle

Deduction, the use of data to test previously developed theories, is represented by the left-hand side of Figure 5.1, while *induction*, the development of theory based on observations, is represented by the right-hand side of Figure 5.1.

The objective of this study as stated in Section 5.1.1 is theory development; this implies that an inductive research method must be adopted. This also suggests that the research philosophy is closer to that of relativism than positivism. However, for the following reasons, a relativist approach is rejected.

1. This study considers many independent/external parameters that affect new product outcomes e.g. market performance, technological developments. Relativists deny this 'external reality' (Robson, 2002).
2. Relativism does not place much value on the ability to generalise beyond the unique situation examined, but for the purpose of this study extending the results beyond the group of firms studied (generalising) is important.

Having rejected both positivism and relativism an alternative research philosophy must be identified. Realism emerges as a suitable approach. It is best described using a diagram, see Figure 5.2.

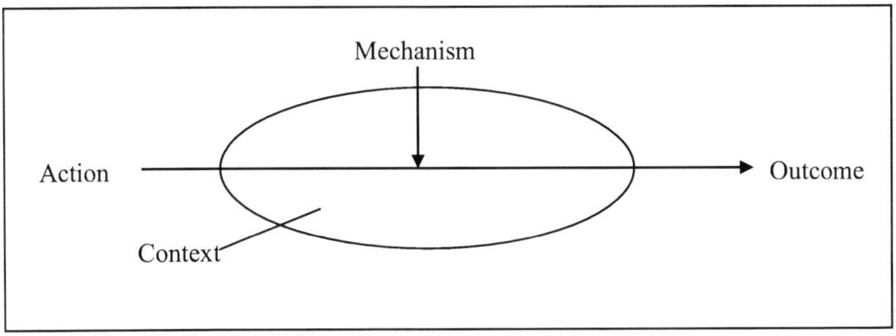

Adapted from Robson, 2002

Figure 5.2 Representation of Realist Explanation

Figure 5.2 illustrates how the outcome of an action is caused through a mechanism within a particular context. In this study the action being examined is the development of new products, the outcome is the level of commercial success. The mechanisms are the NPD management practices and the contexts are the different firm characteristics – size and location. The following points, adapted from Robson (2002) summarise the main features of a realist approach to research.

1. Knowledge is a social and historical product that can be specific to a particular time, culture or situation. No facts are beyond dispute.
2. The task of research is to invent theories to explain the real world and to test these theories by rational criteria. Different methods have to be used for different subject matters.
3. Explanation is concerned with how mechanisms produce outcomes. Whether or not a particular mechanism operates depends on the context.

Having defined the research philosophy to be used in this study, the next section will describe how this philosophy will be applied through the selection of an appropriate research method.

5.3 Research Methods

In keeping with the adoption of a realist research approach, a research method must be selected which is suitable for achieving the research objectives set out in Section 5.1.1. This study is concerned with the management of NPD in small Irish firms but it is being conducted against a background of extensive knowledge of the management of NPD in *large* firms. Additionally, a review of literature on this topic has identified a research project, INTERPROD, that addresses many of the management parameters included in the research model developed for this study. Therefore a logical first step would be the application of the INTERPROD research tool to Irish electronics firms.

The advantages of using INTERPROD are:

1. It is a well-tested research tool that has resulted in a large number of publications. The INTERPROD questionnaires have been developed by internationally recognised experts in the management of NPD.
2. Ireland has not yet been included in the INTERPROD project. Using INTERPROD will allow comparisons to be made between NPD management in Irish firms and NPD management in other countries. In other words, it will position Ireland within an international context.
3. It is very broad-based tool. It examines many aspects of NPD management. It will provide enough data to allow the development of a theory about the management of NPD in small firms.
4. It is a tool that has been shown to be suitable for use with small firms.
5. It is written in English and therefore there should be few problems with interpretation and translation.
6. Access to the INTERPROD methodology and questionnaires as well as data from some other countries participating in the study was made available to the author.

However, despite the advantages listed above, INTERPROD also suffers from some limitations, there are:

1. As with any questionnaire it provides broad but not deep, or rich data (Creswell, 1994).
2. INTEPROD contains a set of questions developed by other researchers and while these do address the NPD management parameters included in the research model they have not been developed specifically to answer the research question that is the subject of this project.
3. Even though INTERPROD has been built on research into NPD management in large firms it has been used successfully on small firms.

4. The success measures used within INTERPROD have known limitations (this will be discussed in more detail later in this chapter).

In the context of this study the advantages of using INTEPROD outweigh its limitations; particularly the ability to compare the Irish data with that from other countries is critical. An international comparison of NPD management in Ireland has not previously been conducted and such a comparison fulfils one of the main objectives of this study.

To overcome some of these limitations, to fully answer the research question and to gain a deeper understanding of NPD in small firms a second research method will also be used. The second method chosen is to develop case studies of three firms and their NPD processes. The reasons for using case studies is to overcome some of the weaknesses of the INTERPROD questionnaire listed above, and also to gain a detailed understanding of how NPD is managed in small Irish firms. The advantages of using case studies in this instance are listed below:

1. Cases will provide deep, rich data that will help to balance the broad data collected by the INTERPROD questionnaires.
2. Surveys and case studies have complementary strengths and weaknesses in terms of research validity and reliability. Therefore, using both methods results in a more robust overall research design.
3. The review of the 13 INTERPROD publications presented in Chapter 3 shows that while INTEPROD is useful in identifying statistically significant relationships within the management of NPD it is not good at providing explanations for these relationships. Case studies on the other hand are particularly useful for explaining relationships and management practices.
4. Developing cases specifically to examine NPD development in small firms can mitigate one of the weaknesses of INTERPROD, that it is based on the results of NPD research conducted in large US firms.

The research plan, as illustrated in Figure 5.3 will consist of two stages; initially INTERPROD will be used to gain a high level understanding of the differences between NPD in small and large firms, then case studies will explore in more detail the issues that emerge from INTERPROD.

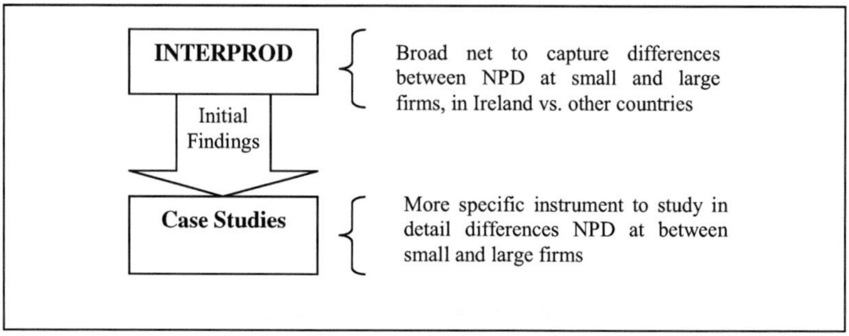

Figure 5.3 Overview of Research Strategy

Using both a survey and case studies means that a multi-method approach has been adopted for this study. The benefits of using multiple methods have been explored by several authors (Robson, 2002; Gill & Johnson, 1997; Brewer & Hunter, 1989; Creswell, 1994). These include:

- Greater validity if both methods reach similar conclusions (see below).
- A better understanding of the concept being explored.
- Minimisation of any bias or weakness inherent in one particular data source.
- Developmental benefits where the first method is used to inform the second method.
- Mixed methods add scope and breadth to a study.
- Reduction of *inappropriate certainty*, a clear cut result from one method may prompt the researcher to think that they have found the 'right' answer, using a second method will encourage more caution.

An additional benefit of using multiple methods is that it facilitates the use of *triangulation.* Triangulation, defined as 'the combination of methodologies in the study of the same phenomenon' (Denzin, 1970), involves examining the same issue from different perspectives. If two, or more, different research approaches arrive at the same conclusion the result is considered to have greater validity (Gill & Johnson, 1997).

A two-phase design, as proposed by Creswell (1994) has been adopted, where two distinct research methods are used sequentially. In this study the quantitative research precedes the qualitative phase. A potential disadvantage of this method is the problem of clearly linking both phases of the study. In this case the strategy for the case studies is developed directly from the results of the survey, clearly linking both phases of the overall thesis. Robson (2002) considers that using quantitative followed by qualitative research methods enhances interpretability – 'in a primarily quantitative study, the interpretation of statistical analysis can be enhanced by a qualitative narrative account'.

A final consideration is the appropriateness of the research methods chosen to the field of study. The management of NPD is often considered within the general area of operations management. The most commonly used method of collecting empirical data in operations management is the survey, followed by the use of case studies (Scudder & Hill, 1998). In other words, both surveys and case studies are commonly used, and considered appropriate, in empirical research in operations management.

The last two research objectives listed in Section 5.1.1 suggest that it is necessary to be able to compare the management of NPD in Irish firms with other countries. This is important for two reasons. Firstly it will position NPD management in Ireland within an international context allowing a discussion about how and why Ireland is different (or not) from other countries. Secondly, it will allow some statement to be made about whether or not the specific characteristics of small firms in Ireland could be expected to apply to small firms in other locations.

This international comparison will happen at two levels. Initially the Irish INTERPROD results will be compared with those from other countries that have been involved in the study and about whom results have been published, this will be presented in Chapter 6. Secondly, to explore at a deeper level the extent to which NPD management in Ireland is unique (or otherwise) a more detailed examination of Irish and UK data will be presented in Chapters 7 and 8.

The UK was selected for this detailed comparison for three reasons. Firstly, the UK is a close trading partner, both countries speak English, and there is a high level of flow of people, goods, and technology between the UK and Ireland. It could be expected that electronics firms in both countries operate under similar conditions. Therefore, any differences between the UK and Irish data can be attributed to some difference in the Irish situation. Secondly, at a pragmatic level, links between the author and researchers in the UK provided access to the UK INTERPROD data. Finally, there is a critical mass of research into the management of NPD in the UK. Comparing Irish data with data from the UK will allow the author to draw on the findings of this research in an attempt to understand the management of NPD in Ireland.

The previous discussion about comparing the Irish INTERPRDO data with the UK and the 'rest-of-the-world' suggests that the primary research question should be developed in recognition of the potential insights to be gained through international comparisons. Therefore the research question can now be stated as:

1. *How do small Irish electronics firms manage the development of new products, and are the success factors that apply to small Irish firms different from those that apply to large Irish firms?*
    - 1.a *Are the differences between small and large Irish electronics firms also seen between small and large UK electronics firms?*
    - 1.b *How does NPD management in the Irish electronics sector, as measured by the INTERPROD research tool, compare with that in the UK?*

It is now also possible to present the research plan detailing what method is used, the objective of each stage and the chapter in which results are presented and discussed, see Table 5.1.

Research Method	Objective	Chapter(s)
Survey – INTERPROD	Comparison of Irish INTERPROD data with INTERPROD data published about other countries in order to *explore* the management of NPD in Irish firms.	6
Survey – INTERPROD	In depth comparison of Irish and UK INTERPROD data, *exploring* the differences between small and large firms in each country and *describing* the management of NPD in Irish firms.	7 and 8
Case Studies	Detailed examination of NPD management in small and large Irish firms using case studies, to *explain* some of the relationships identified by the survey and to further *describe* NPD management in Irish firms.	9 and 10

Table 5.1 Research Plan

The next two sections in this chapter will discuss the use of the INPERPROD survey and the case studies respectively.

5.4 INTERPROD

Having identified INTERPROD as a suitable research tool for this study the researcher met with Dr William Souder, The University of Alabama in Huntsville, USA, the director of the INTERPROD project. Although the project was nearing completion it was agreed that the author would administer the INTERPROD questionnaires in Ireland. This brought the number of countries involved in INTERPROD to fifteen.

INTERPROD is a very broad research instrument; it asks 214 questions at company level and 156 questions at project level. It covers a wide rage of issues including organisational issues, development process factors, project life cycle, portfolio management, and market and customer characteristics. Additionally, as shown in Table 5.2, all twelve variables included in the research model are addressed within the INTERPROD survey. From this perspective it was considered suitable to apply to the first phase of this project – it provided an overview of the differences between NPD management in small and large firms.

Area of Analysis	Unit of Analysis	Questionnaire
New Product Characteristics	Project	INT-C
Marketing Characteristics	Project	INT-C
Marketing and Technological Synergy	Project	INT-C, INT-D
Resources for NPD	Company	INT-A
Top Management Support	Project	INT-B
Marketing and Technological Skills	Project	INT-C, INT-D, INT-E
Development Process Proficiency	Project	INT-B, INT-L
Project Manager Skills	Project	INT-B, INT-E
Cycle-time Practices	Company	INT-A, INT-F
Involvement of External Organisations	Company	INT-A
R&D/Marketing Integration	Project	INT-B
Methods of Organising NPD	Project	INT-B

Table 5.2 Areas Analysed within the INTERPROD Data

INTERPROD also has the advantage of examining new product development at both the project, or micro, level and the company, or macro, level. Few studies have taken this broad approach. The advantages of including a company level analysis, in addition to project level analysis, are detailed by Cooper & Kleinschmidt (1995).

1. In studying NPD success factors examining only project level success may be limiting. While successful new product projects are obviously an advantage to a firm, a series of small incremental new product projects may not necessarily lead to success at the company level.
2. The importance of some company level practices, for example having a clear technology strategy, will not be apparent when only project level practices are observed.

3. When examining success-failure project pairs within firms, company level practices are common to both projects and therefore will not emerge as success factors during analysis.

The rest of this section will describe how the INTRPROD data were collected and also how they were analysed.

5.4.1 Collecting INTERPROD Data

The manner in which the INTERPROD data were collected varied from country to country. In the US researchers worked closely with the firms involved, each firm was asked to report on 2 successful and 2 unsuccessful projects with an even number of projects completed under conditions of high and low market and technical uncertainty. The number of projects describe by each firm in the US is quite high with firms reporting on an average of 3 projects. In the UK the questionnaires were sent by post and though the number of firms involved was high, 63, the number of projects per firm was less than two. Data collected in Ireland lay somewhere between these two cases.

Data in most countries were collected across a range of high-technology industries; the criteria for inclusion in the study were that firms operate in an area of high technology and also that they supply products to industrial customers. The Irish researcher was only interested in the electronics sector so all the Irish firms involved in the study operated within the electronics sector.

As this study examines UK and Irish INTERPROD data in detail, this section will examine how data was collected in these two countries.

Irish INTERPROD Data

The firms involved in this study all operate in the electronics sector, developing and manufacturing a variety of products from integrated systems to discrete components. Most of the data presented in this report was collected by the author during the summer of 1998. Three firms were involved in piloting the research tool during 1997; data from these three firms has also been included in the report. While not all firms, in either the group of large or small firms, were Irish-owned, they all have Irish management and were responsible for developing new products from the exploratory stage through to providing after sales support and service.

The development of 63 new products in 36 electronics firms was examined; 14 of the firms studied were large (100 or more employees); the remaining 22 were small (fewer than 100 employees).

Companies for inclusion in the study were selected from an on-line Kompass data-base (1997). These are described in Table 5.3 and are mapped to the closest applicable NACE and SIC codes.

Sector Description	**NACE Code**	**Corresponding SIC Codes**
Manufacture of office machinery and computers.	NACE 30	3571, 3695
Manufacture of electric motors, generators and transformers, electricity distribution and control apparatus, as well as insulated wires and cables, accumulators, batteries and lighting equipment.	NACE 31	3823, 3825
Manufacture of electronics components, television, radio and video equipment, as well as telephones.	NACE 32	3661, 3663
Manufacture of medical, precision and optical instruments, watches and clocks.	NACE 33	3845

Table 5.3 Industrial Sectors included in Irish INTERPROD study

A total of 882 companies were selected. This list was further refined eliminating companies that were obviously not electronic e.g. dental labs, and those that were known to the author not to be involved in R&D e.g. electronics sub-contract assembly, cable and harness assembly etc. Of the remaining 201 firms, a further 5 were eliminated as, on further investigation, they were not involved in electronics; this left 196 firms to be contacted by phone. An attempt was made to contact each of these firms by phone - 14 firms could not be contacted; the phone numbers were either incorrect, the firms were no longer listed in the telephone directory or the numbers rang out. The R&D managers in the remaining 182 firms were contacted by phone. Based on the initial phone contact, firms were divided into five categories as shown in Table 5.4.

Description	Number of Firms
Not Electronics (predominantly S/W only),	47 (26%)
Electronics R&D and Manufacturing,	56 (31%)
Electronics R&D but not the complete process	7 (4%)
Minimum Electronics R&D (only 1 or 2 people, no formal department)	8 (4%)
Electronics but no R&D	64 (35%)
Total	**182**

Table 5.4 Results of Initial Contact with Firms

For the purpose of this study the valid sample consists of the 56 firms in category 2 above, i.e. those firms with the complete R&D process, from product specification through to commercialisation.

Five attempts were made to speak with the R&D manager/managing director. If contact could not be made the firm was considered to be non-respondent. Those that could be contacted and agreed to proceed further were then sent an information pack a (see Appendix 3). A total of 53 information packs were sent; however some packs were sent to firms who were subsequently found to be unsuitable. A total of 36 firms agreed to be interviewed, this constituted a response rate of 64%.

At the initial meeting, which normally lasted 1 - 2 hours, as many questionnaires as possible were completed. In some cases the MD or R&D manager felt qualified to complete all questionnaires; this was most likely to occur in smaller firms, while in other cases additional people within the company were identified and subsequent meetings were organised. Some concern could be expressed about using only a single respondent; however, where the respondent 'possess unique process insights', (Cooper & Kleinschmidt, 1995), this approach is considered valid.

All firms were requested to discuss two projects - one that had been a success and another that had been a failure. The decision on success and failure was left to the managers being interviewed. Not all firms were able to identify both successful and unsuccessful products and in some cases only one NPD project was examined. Table 5.5 summarises the Irish sample in terms of firms and projects.

	Large Firms	Small Firms	Total
Number of Firms	14	22	36
Number of Successful Projects	15	20	35
Number of Unsuccessful Projects	11	17	28
Total number of projects	26	37	63

Note: Small firms are defined as those with fewer than 100 employees.

Table 5.5 Breakdown of Irish sample

As can be seen in Table 5.5 the Irish data is evenly balanced in terms of successful and unsuccessful products, in both small and large firms. Successful projects make up 54% of the sample from small firms and 58% of the sample from large firms. Additionally, as the focus of the Irish study is on small firms and the differences between these small and large firms, the Irish sample has been constructed to include approximately two thirds of small firms and a third of large firms.

UK INTERPROD Data

The INTERPROD data was collected in the UK by Prof. Rachel Cooper and Joanne Charlton. A total of 63 high technology firms participated in the study. These came from a variety of manufacturing sectors including electronics, instruments, software, pharmaceuticals, food, chemicals and fibres and fabricated products. Companies were identified through business directories and were contacted by letter. In total, 656 UK companies were contacted and 63 useable responses were received, a response rate of 9.6%. The data was collected during 1994 – 1996, about 2 years earlier than the Irish data.

INTERPROD data was collected from a range of high technology industries in the UK. Of the 63 firms involved in the study 25 were involved in the electronics industry. A decision had to be taken about whether to include all 63 UK firms in the detailed comparison between UK and Irish firms or whether to limit the UK sample to the 25 electronics firms. Using data from all 63 firms would have increased the sample size and thus the statistical validity of the results; however if electronics firms were found to be significantly different from the rest of the UK firms, then the use of non-electronics firms would have resulted in invalid comparisons.

In order to decide whether or not to examine the complete UK sample the electronics firms were compared with the rest of the sample. This preliminary examination shows several significant differences between the electronics firms and other UK firms involved in the study; some of these are shown in Table 5.6.

Parameter	All UK Firms[1]	UK Electronics Firms[1]	Comparison of Medians
% Turnover from new products	32.5%	60%	.012[2]
% Employees in R&D	8%	21%	.007[2]
R&D expenditure as a % of sales	7%	11%	.003[2]
Spending on Process Improvement	10%	7.5%	.027[2]
Spending on Cost Reduction	6.5%	5%	.032[2]
% Time spent on prototype development	20%	30%	.001[2]
% Time spent on manufacturing start-up	10%	5%	<.0005[2]

Note 1: The figures given below are medians

Note 2: Statistically significant, $p \leq 0.05$, Mann Whitney U test

Table 5.6 Differences between UK electronics and UK non-electronics firms.

This implies that electronics firms are significantly different from other high technology firms in terms of the amount that they invest in NPD, how they allocate NPD resources, and how their NPD processes are executed. These differences are not surprising; the nature of R&D, the role of technology and the market environment differs greatly from one industry sector to another (Hoffman et al., 1998), and they demonstrate that for the purpose of this study it is better to compare only the 25 UK electronics firms with the sample of Irish electronics firms.

The questionnaires used in the UK were identical to those used in Ireland and in the 14 other countries involved in this study. Using the same questionnaire in different countries has facilitated many comparative studies (Balbontin, et. al.1999, Souder et. al. 1997, Souder & Jenessen, 1999, Souder et. al. 1998, Song et. al. 1997, Souder & Song, 1997). The only difference between the Irish and UK questionnaires was the inclusion of a small number of additional questions in the Irish questionnaires; these were all company-level (rather than project-level) questions and are discussed in Chapter 8.

Five of the 25 UK electronics firms were small while the remaining 20 were large. The breakdown of the UK electronics firms is given in Table 5.7.

	Large Firms	Small Firms	Total
Number of Firms	20	5	25
Number of Successful Projects	21	6	27
Number of Unsuccessful Projects	8	4	12
Total number of projects	29	10	39

Note: Small firms are defined as those with fewer than 100 employees.

Table 5.7 Breakdown of UK Electronics sample

The UK sample is not balanced in terms of successful and unsuccessful projects in small and large firms. Successful projects make up 60% of the sample from small firms and 72% of the sample from large firms. Additionally, the UK data was not collected in order to examine small firms; small firms only make up 20% of the UK sample.

The fact that there are only five small firms within the UK INTERPROD data will limit the extent to which small firms in Ireland and the UK can be compared. However, the usefulness of the UK data is as follows:

a. To compare the overall results from Ireland and the UK and
b. To compare large UK and Irish firms.

It may be possible to make some comment about NPD management in small Irish and small UK firms but it is unlikely that this comparison will result in many statistically significant results.

The reasoning that will be used in discussing these results involves examining the differences between 1) small Irish and large Irish firms, 2) large Irish and large UK firms and 3) Irish and UK firms. This approach will provide insights into the management of NPD in Irish firms (compared to UK firms) and then, within this group, it will facilitate better understanding of practices in small Irish firms (through comparison with large Irish firms).

5.4.2 Definition of Success within INTERPROD

As described in Chapter 3, firms participating in the INTERPROD project were required to report on one successful and one unsuccessful NPD project, and the management practices employed in both projects were then compared. Two measures of success were collected; respondents were asked to describe the 'commercial outcome' and the 'technical outcome' of each project and the extent to which the anticipated outcomes were realised. The exact format of these questions can be seen in INT-B in Appendix 1. Commercial success as defined in previous INTERPROD publications will be used within this study and is described briefly below.

Each project was ranked based on how well it met commercial expectations. Table 5.8 shows the rating system used. Failed new products were those that scored lower than 3 on commercial outcomes while successful projects met or exceeded the expectations of the firm.

Score	Commercial outcome of the product
1	Far below our expectations
2	Slightly below our expectations
3	Consistent with our expectations
4	Slightly above our expectations
5	Far above our expectations

Table 5.8 Project outcome scoring system

Technical success was measured in the same way, except that respondents were asked to indicate the degree to which the technical outcome of the project met or exceeded the expectations of the firm.

Section 4.3 described some of the difficulties in measuring new product success. Hart (1993) examined this issue in detail. She reported that there is little consistency within the literature in NPD about what constitutes success; a range of financial and non-financial measures are used. Based on a survey of 69 British manufacturing firms Hart attempted to determine which measures of success were most appropriate when studying NPD management. She found few significant relationships between financial and non-financial measures of success and concluded that success is multi-dimensional.

Given that there is not a universally-agreed definition of new product success within NPD literature a decision had to be made about the success measure to be used in this study. It would be possible to measure success based on a range of financial and non-financial outcomes. However, the uniqueness of such a measure would limit the ability to make comparisons with previous research, particularly INTEPROD studies. The measure used in previous INTERPROD studies is limited but has the advantage of facilitating international comparisons. The INTERROD papers published to date have used projects' commercial outcomes as a measure of their success. This approach is limited for two reasons: 1) it does not consider non-commercial success (e.g. projects that have resulted in valuable technical learning) and 2) it is subjective (the respondent decides if the project is a success without comparing actual performance between projects).

There is very little published to date about the management of NPD in Irish firms. One of the objectives of this study is to explore the extent to which the management of NPD in Ireland is different from that in other countries. Therefore, the ability to compare the results of Irish NPD management data with that from other countries is critical to this study.

5.4.3 Analysing INTERPROD Data

This section examines how the data was analysed, what statistical methods were used and how they were applied.

Data Analysis Methods

The data was analysed in two different ways. Where the data were appropriate (on the *interval* or *ratio scales*) parametric methods (i.e. the calculation of means, variances etc.) could be applied, in other cases non-parametric methods were used. For example, most of the project level data explored the use of different management practices, this data was recorded using an *ordinal scale* and was correlated (using Spearman rank correlation) with the extent to which a project met or exceeded its commercial expectations. Spearman correlation coefficients (Siegel and Castellan, 1988) were calculated comparing the project outcomes, as shown in Table 5.8 above, and the responses to a range of questions about the management of each project. This allowed management practices that were linked with successful new products to be identified.

The Spearman rank-correlation coefficient has been developed to measure the association between two variables when only ordinal data are available. In the case of much of the INTERPROD data the level of project success (measured on an ordinal scale) is compared with the degree of application of a management practice (again measured on an ordinal scale). The Spearman correlation coefficient is appropriate to analysing this type of data.

Concerns about the reliability of the INTERPROD research instrument relate to the possibility of significant relationships between independent and dependent variables being overlooked – in other words Type II errors. This study is exploratory; it is concerned with identifying possible relationships between firm size and successful NPD management practices. Relationships identified will be subjected to further examination or proposed as areas for future research. In this context it is more serious if actual relationships are overlooked rather than if spurious relationships are further examined. Type II errors can be reduced by using a higher level of significance than the normally accepted 0.05 and 0.01; therefore, within Chapter 7 and 8 results that are significant up to a level of 0.2 will be highlighted. Using a higher significance level will be particularly useful when analysing the small UK electronics firms – it is difficult to find significant relationships when using small sample sizes. This issue will be discussed on more detail in Chapter 7.

5.5 Case Studies

Yin (1989) defines three conditions that help a researcher to select an appropriate research strategy:

1. The form of the research question.
2. The control that the researcher has over behavioural events.
3. The degree of focus on contemporary events.

Where the research questions are in the form of *how* or *why*, the researcher has no control over events. When the events being studied are contemporary, the use of case studies is a suitable strategy. All three of these conditions apply to the final stage of this research study. The case studies are being used to get a deeper understanding of *how* NPD is managed in small firms and also to understand *why* small firms are different from large firms; also, the researcher has no control over the contemporary events being studied.

Yin (1989) also provides a 'technical' definition of a case study as being an empirical inquiry that:

- *investigates a contemporary phenomenon within its real-life context; when*
- *the boundaries between phenomenon and context are not clearly evident; and in which*
- *multiple sources of evidence are used.*

Again the case studies used within this research project meet this definition.

Case studies have been criticised as being a 'soft-option' and having 'dubious value'; by themselves (Robson, 2002). However, such criticism often stems from an unclear understanding of the role of case studies. Case studies are not intended to generalise theories to populations. Rather, they are used to develop 'theoretical propositions' (Yin, 1989). Within this project the role of the three case studies is not to prove any of the relationships presented in the research mode, but to gain a better understanding of the nature of these relationships. As discussed above, this study is about theory building therefore case studies are an appropriate research strategy.

Yin (1989) defines three different applications for case studies. Firstly, they can be used to *explain* causal links that are too complex for other strategies. Secondly, they can be used to *describe* a real-life context. And, finally they can be used to *explore* different situations. Specific cases do not have to fulfil all three applications. In this study it is hoped that the case studies will be able to:

- Explain the links between some management practices and new product success,
- Describe the management of NPD at small firms, and,
- Explore NPD management in three different situations.

The analysis of the cases will further inform the research model and provide a better understanding of the impact of firm size on NPD management.

The rest of this section will describe the way in which case study data was collected and analysed.

5.5.1 Collecting Case Study Data

The Irish INTERPROD sample was structured to contain approximately two small firms for each large firm. This ratio of small to large firms will be extended to the case studies so that three case studies will be conducted, one based in a large firm and two in small firms. Cases were selected to reflect the diversity of the processes used to develop new products in small and large firms. The three firms involved in the study are summarised in Table 5.9. The number of interviews conducted in each firm increased with firm size and was large enough to include all key areas within each firm's NPD function. All three firms operate within the electronics industry and are located in the mid-west region of Ireland.

Telephase Inc.	**Ferrostar Ltd.**	**eFlex Systems Ltd.**
1,200 employees	90 employees	9 employees
Interviewees x 13 Vice President x 1 Team Leaders x 4 Developers x 7 New Product Co-Coordinator x 1	*Interviewees x 9* Managing Director x1 Development Manager x 1 Marketing Manager x1 Developers x 4 Technician x 1	*Interviewees x 5* Managing Director x 1 Engineering Manager x1 Marketing Manager x1 Development Engineers x1 External Consultant x1
9 Product Lines 1 Business Site	2 Product Lines 1 Business Site	1 Product Line 1 Business Site

Table 5.9 Summary of Case Study Firms

Initial analysis of the INTERPROD data suggested that small firms cannot be considered to be a homogenous groups and that very small firms, i.e. fewer than 20 employees, are different from small firms with 20 to 100 employees. Therefore, the small firms were selected to reflect these different groups.

Many of the criticisms of case studies related to research being conducted in a 'sloppy' manner (Yin, 1989; Robson, 2002). However, this criticism could equally be levelled against other research strategies, where many surveys are poorly constructed and analysed. Yin defines three principles of data collection that help to increase the validity and reliability of a case study. The application of these principles within this research project is described below:

1. **Using Multiple Sources of Evidence:** The case studies presented rely on two main sources of evidence, interviews and documentation. The documentation in each case consists of the firm's NPD procedure and any supporting documents. The interviews were conducted with a number of key individuals involved in the development of new products including senior management.
2. **Creating a Case Study Data Base:** Data waws collected using interviews that lasted 1 – 1.5 hours. The interviews were guided by a series of questions -see Appendix 4- but they were open ended in the sense that respondents were encouraged to express opinions about how NPD was managed within their organisation. The interviews were recorded manually by the researcher and later transferred to computer. In all cases the typed interview was e-mailed back to the respondent asking then to check its accuracy and, if necessary, to add further comments. For each firm, a file has been set up containing the original interview questions, the hand-written and typed questionnaires, and any other documentation (procedures, brochures etc.) relating to the firm.
3. **Maintaining a Chain of Evidence:** A chain of evidence in a case study allows a reader to relate the conclusions back to the evidence originally gathered. The chain of evidence in this project is built by making extensive use of direct quotations in presenting the case material. This would allow an 'investigator' to move from the case reports presented in Chapter 8 back to the original interview material identifying the respondent, time and location of the interview. The chain would be easier to establish if respondents were named under each citation in Chapter 9; however, for reasons of confidentiality this has not been done. Additionally, a chain of evidence has been maintained between the INTEPROD data and the case studies. All three case study firms participated in INTERPROD and the initial contact in each firm was with a person who had also completed some of the

INTERPROD questionnaires. Many of the people who took part in the case study interviews had also taken part in INTERPROD.

Adhering to the principles listed above increases the reliability and construct validity of the case studies, (Yin, 1989) thereby facilitating analysis of the cases. The method used to analyse these case studies is described below.

5.5.2 Analysing Case Study Data

Two general strategies have been defined for analysing case studies, (a) developing a case description and (b) exploring theoretical propositions. Because this study has a guiding model and since the case studies were developed to explore in more detail some of the results of the INTERPROD survey, there are several theoretical propositions that can be used in analysing the case studies. The advantage of such a strategy is that the theoretical proposition helps to focus the case studies on the most relevant data. This method can be described as a 'Prestructured Case' (Miles & Huberman, 1994). Using a prestructured case has the advantage that it helps to focus and streamline the collection of data but a disadvantage is that it can lead the researcher to early conclusion drawing. One way to avoid this disadvantage is to use other forms of information in addition to the case material; in this case INTERPROD has provided that other source information.

The case studies are analysed in Chapter 10, the analysis structured into the same five areas that guided the presentation of the cases:

1. Cycle-time practices
2. Resources for NPD
3. Methods of organising for NPD
4. Involvement of external organisations
5. Project management skills.

A justification for the selection of these five headings, in the light of the findings from INTERPROD, is given at the end of Chapter 8.

There are three dominant modes of analysing case studies: pattern matching, explanation building, and time series (Yin, 1989). The second of these, explanation building, is used in analysing the cases in Chapter 10. Yin suggests that explanation building is used primarily in analysing explanatory case studies that are part of a hypothesis (or theory) building process. These cases are used to *describe* how NPD projects are managed in small firms. They are also intended to *explain* some of the causal links identified by INTERPROD.

The case studies were written by examining the contributions from all of those interviewed under the five headings listed above. Comments from each respondent and as relating to each issue were grouped; these were then used to describe the management of NPD within each area studied. The case studies presented in Chapter 9 include a large number of direct quotations. In writing the cases based on the smaller firms it was possible to include comments from most interviewees. However, in the case of the large firm this would have resulted in a lot of repetition and (therefore) representative comments were selected. The overview of each firm was based on the initial contact; in the two small firms this was the R&D manager; in the large firm the main contact was the NPD process manager. In all cases the researcher was provided with a copy of the NPD procedures and an NPD organisation structure. These were described in detail at an early meeting and helped in the selection of interview subjects. Throughout the collection of data and during the writing of the cases, close contact was maintained with the person who was the initial contact within the firm. A version of the case study was provided for each firm for comment and discussion. Any relevant feedback was included in the final version of the case.

Several techniques were used for theory-building (Miles & Huberman, 1994) and are listed below along with an example of how each was employed.

1. **Noting Patterns, Themes:** This involves noting recurring patterns or themes that can be used to pull together separate pieces of data. For example, within eFlex there was a sense from all those interviewed that they worked as a team, and that they had moved from a situation that was out of control to one where all staff were working towards the same goals.
2. **Seeing Plausibility:** When analysing cases it is possible to draw some conclusions based on observations that are intuitive or 'make good sense'. This approach was used in the eFlex cases where the benefits that the firm was gaining from its collaboration with the external technical consultant were described during the interviews with the consultant, the engineering manager and the managing director. In Ferrostar, the

difficulty that was experienced with using external technical consultants was related by three of the technical managers interviewed.

3. **Counting:** Case study data can be analysed by counting the occurrences of particular responses or comments. For instance, in the large firm Telephase, it was possible to count some responses. Everyone interviewed was asked about controlling NPD costs; none of the engineers or managers interviewed could describe how NPD costs were controlled but several referred to managing time.
4. **Making Contrasts/Comparisons:** Making comparisons between different sets of data is a well-established method of drawing conclusions. For example, within Telephase two of the NPD teams were using fulltime project managers. It was possible to contrast the comments that came from engineers in these teams with those working in other teams. Those working with fulltime project managers were positive about their ability to deliver NPD projects on time while those in other groups were not.
5. **Building a Logical Chain of Evidence:** In analysing case studies, separate pieces of data can be brought together to illustrate a point. For instance, in the Ferrostar case several of those interviewed commented on how the firm used to be run when it had fewer employees and how, despite several initiatives, the NPD process was now not running smoothly. The managing director also described the firm as being stuck in the past. This allowed a theory to be developed about the difficulties experienced by firms as they grow to be developed, and as supported by evidence from INTERPROD.

5.6 Evaluation of Research Design

Any method of research has an inherent set of strengths and weaknesses (Gill & Johnson, 1997). There are two important characteristics that can be used to evaluate a research method (Tiernan, 1995):

1. *Validity* relates to the degree to which the research measures what it was intended to measure. It can be further divided into interval validity and external validity:

a. *Internal Validity* is a measure of the extent to which cause and effect relationships identified by the research are in fact valid.
b. *External Validity* refers to the extent that these cause and effect relationships can be extended beyond the data set and the setting involved in the study. External validity has two components:
    i. *Population Validity* refers to the possibility of generalising from the sample to the wider population.
    ii. *Ecological Validity* refers to the extent that it is possible to generalise from the context or setting for the research to other contexts and settings.

2. *Reliability* is concerned with the consistency of the results obtained. In order to satisfy the need for reliability it must be possible for another researcher to obtain the same results by repeating the original research with the same respondent and conditions.

The strengths and weaknesses of the different research methods used are summarised in Table 5.10.

	Internal Validity	Population Validity	Ecological Validity	Reliability
INTERPROD Survey	Weak	Strong	Medium	Strong
Company Case Studies	Strong	Weak	Weak	Strong

Table 5.10 Evaluation of Research Methods

There now follows a brief description of INTERPROD and the case studies under the four headings defined above: internal validity, population validity, ecological validity and reliability.

INTERPROD

Internal Validity: Analysis of the INTERPROD data to establish causal relationships is carried out using correlation, but while correlation will detect that two variables change in the same way, it does not provide sufficient proof of a causal relationship; in other words, the internal validity of a survey, i.e. INTERPROD, is low.

Population Validity: The INTERPROD study included Irish electronics firms that were involved in developing new products with a substantial hardware element and who were also involved in manufacturing. The response rate was 64% of the target population therefore the validity of making inferences about the complete population based on the sample involved in the study is good.

Ecological Validity: Since Irish firms developing electronics hardware from a small and specialised group, and, as shown above, NPD varies significantly from industry to industry, the ecological validity of the INTERPROD survey in Ireland in not very strong. The advantage of using the INTERPROD survey, however, is that it has been conducted in many different countries, and on many different types for firms; therefore, the findings from the Irish firms can be compared with a larger database of results. This increases the ecological validity of the study.

Reliability: The INTERPROD questionnaire has been used in 19 different countries and has yielded comparable results from each country. Many of the findings of the survey have been consistent with other studies and have been widely published. The fact that there were consistent results across many measures between the Irish and the UK results supports the argument that the survey is reliable. A review of the publications based on INTERPROD further strengthens this argument.

Case Studies

Internal Validity: In contrast to surveys, case studies have strong internal validity. The descriptive nature of the cases allows relationships to be observed in considerable detail and the causal relationships observed by INTERPROD to be explained.

Population Validity: The purpose of the case study is not to generalise to the wider population and it cannot be used for this purpose. The cases are used to provide explanations for relationships identified by INTERPROD and to build theory, not test it.

Ecological Validity: As above, ecological validity does not apply to case studies; this measure is more appropriate when using other research methods.

Reliability: By adhering to the three principles of data collection the reliability of the case studies is improved. In other words, using multiple sources of information, controlling documentation and maintaining a chain of evidence all make it possible for the findings of these three cases to be repeated by another researcher.

The above analysis demonstrates the advantages of using a multi-method research approach. The weaknesses in the survey approach are balanced by strengths in the case studies. The main weakness identified in the study is the ability to generalise beyond the setting of Irish (and UK) electronics firms and, as described above, the use of INTERPROD helps to overcome this weakness.

5.7 Conclusion

This chapter has provided an overview of the research plan to be used in this volume – a multi-method approach employing a quantitative survey and a qualitative case study. The objectives of the research strategy were described in the chapter introduction. This was followed by a brief discussion on research philosophy that concluded with the selection of a realist approach. The two research methods to be used were identified and the advantages of using each method presented. The research question presented in Chapter 1 was extended to explore the extent to which NPD management in Irish firms was different from that in other countries.

The two research methods, i.e. the INTERPROD survey and the case studies were then described in terms of how the data was collected and analysed. Finally the strengths and weaknesses of each approach was presented. While there are weaknesses associated with using either surveys or case studies in isolation this project utilised a multi-method strategy in which the weaknesses of one method are balanced by the strengths of the other.

CHAPTER SIX

INTERPROD RESULTS AND DISCUSSION

6.1 Introduction

This chapter presents the initial analysis of the Irish INTERPROD data. The purpose of this chapter is to develop an understanding of where Ireland fits into the complete INTERPROD data set. The simplest way to achieve this is to replicate the analysis that has already been conducted on the INTERPROD data, as presented in Chapter 3, and as using the Irish data. However, as discussed in the conclusions of Chapter 3, there is a lack of consistency about the way that the INTERPROD data has been analysed and there is also considerable overlap between the analyses in many of the publications. Therefore, the Irish data will only be compared directly with the results from five INTERPROD publications that are considered most relevant to the Irish situation. The rationale behind the selection of these five publications is presented in Section 6.1.1 below.

The rest of this chapter is laid out as follows; Section 6.2 presents the results based on the analysis of the Irish data; Section 6.3 contains a discussion of these results and Section 6.4 re-examines the research model for this study in the light of the data presented in this chapter. The conclusion of this chapter is in Section 6.5.

The key findings of the analysis and discussion contained in this chapter are:

1. There are few NPD success strategies that apply in all situations – the successful management of NPD is very context-dependent.
2. Irish firms are involved mainly in 'regular' or incremental innovation projects.
3. The management of NPD in Irish electronics firms is influenced by a strong US presence within the electronics sector in Ireland

4. Several factors, including organisational factors, market parameters, and technical skills were not linked with new product success in Irish firms.

6.1.1 Key INTERPROD Publications

The focus of this volume is on small Irish electronics firms. It is useful in examining this group of firms to compare the Irish INTERPROD data with other studies that included small firms. Ireland has a small open economy; it has frequently been compared with New Zealand and Korea (STIAC, 1995). Comparing the Irish data with data from these, and other similar economies, will provide insights into the management of NPD in Ireland.

In order to streamline the analysis of the Irish data and to allow for direct comparison with the most relevant of the published INTERPROD papers, the Irish data will be analysed under the headings shown in Table 6.1. The list shown in Table 6.1 was selected to include the publications that (1) considered firm size as a variable, (2) were based on countries with small open economies (New Zealand, Korea, Belgium), and (3) addressed variables included in the proposed research model presented in Figure 4.1. Question codes have been placed in square brackets after each of the variables listed in Table 6.1. These make it possible for the reader to locate the corresponding questions within the seven INTERPROD questionnaires in Appendix 1.

The five publications that reported on the variables listed in Table 6.1 are:

1. Yap & Souder, 1994; Factors influencing new product success and failure in small entrepreneurial high-technology electronics firms.
2. Souder & Song, 1997; Contingent product design and marketing strategies influencing new product success and failure in US and Japanese electronics firms.
3. Souder, Buisson & Garret, 1997; Success through customer driven new product development: A comparison of US and New Zealand small entrepreneurial high-technology firms
4. Lee, Lee & Souder, 2000; Differences of organisational characteristics in new product development: cross-cultural comparison of Korea and the US
5. Wren, Souder & Berkowitz, 2000; Market orientation and new product development in global industrial firms.

The list of variables in Table 6.1 has been rationalised in order to avoid duplication. The list provides a comprehensive review of the INTERPROD data, it also offers a wide range of coverage and will allow for meaningful comparisons to be made between the Irish data and that presented in previously published INTERPROD papers.

1. Entry Strategies: Source: Yap & Souder, 1994	Positioning: Against another product [C28] Positioning: Product class dissociation [C29] Positioning: Benefits and solutions emphasised [C25] Distribution: Developers own salesforce [C30] Distribution: Manufacturers representatives [C32]
2. Performance superiority: Source: Souder & Song, 1997	Product positioned on specific product features [C24] Product positioned on the benefits and solutions [C25] Superior Performance as a planned selling point [C38]
3. Technical superiority: Source: Souder & Song, 1997	Technical Superiority as a planned selling point [C39]
4. Radicalness: Source: Souder & Song, 1997	Unique Features were a planned selling point [C40] Product positioned on product class dissociation [C29] Users required to change way of doing things. [C22]
5. Compatibility: Source: Souder & Song, 1997	Product Compatibility as a planned selling point [C46] Product positioned on specific usage occasions. [C26][a]
6. Users eagerness to adopt: Source: Souder & Song, 1997	Potential customer had a great need for product. [C8] Customers were satisfied with existing products. [C12][a] Purchase of this product was major decision [C10][a]
7. Market growth potential: Source: Souder & Song, 1997	Market for this product was growing quickly. [C14] The window of opportunity was big. [C18] The users' needs change quickly in this market. [C9] The market for this product was mature [C15][a]
8. Intensity of competition: Source: Souder & Song, 1997	The market for this product was competitive. [C16] There were many competitors in this market. [C17]
9. Market proficiency: Source: Souder & Buisson & Garret, 1997	Market Development activities [B54] Market Research [C48] Marketing Start-Up activities [B56] Sales and Marketing activities [C49]
10. Technical Proficiency: Source: Souder &	Exploratory activities [B50] Concept Development activities [B51] Prototype Development activities [B52]

Buisson & Garret, 1997	Prototype Testing activities [B53] Manufacturing Start-Up activities [B55] Technical Service activities [B57]
11. Technical skills: Source: Souder & Buisson & Garret, 1997	R&D skills were at the desired level [D10] Engineering skills were at the desired level [D11] Manufacturing skills were at the desired level [D12] Perceived technical expertise was very high [E9]
12. Marketing skills: Source: Souder & Buisson & Garret, 1997	Forecast of market demand was accurate [C50] Predictions of cust. requirements were accurate [C51] Close fit with marketing skills [C54]
13. Project management: Source: Souder & Buisson & Garret, 1997	Project manager had technical skills [B28] Project manager had marketing skills [B29] Project manager had management skills [B30] Project manager had participative style [B31] Project manager was able to motivate [B32]
14. R&D/marketing integration: Source: Souder & Buisson & Garret, 1997	Level of contact between commercial and technical [B18] Level of information flow [B19] Participation by commercial in problem definition [B21] Participation by technical in problem definition [B22]
15. Organisation for NPD: Source: Lee, Lee & Souder, 2000	How was the project organised and directed? [B17]
16. Organisational processes: Source: Lee, Lee & Souder, 2000	Level of procedure formality [B13] Level of authority concentration [B14] Level of participation in decisions [B15] Degree of organicity [B16] R&D/Marketing integration [B18 – B22]
17. Customer orientation: Source: Wren, Souder & Berkowitz 2000	Cust. problems with previous products considered [C23] Customer's needs were well defined [C6] Forecast of market demand was accurate [C50] User requirements well understood [E6]
18. Top management support: Source: Wren, Souder & Berkowitz 2000	Top management was supportive of the project [B23] Top management involvement in the project [E8]

[a] These items have been reversed in the statistical calculations.

Table 6.1 Factors to be analysed in the Irish INTERPROD data (continued)

The eighteen variables included in Table 6.1 all address project level variables and can be mapped onto the project level variables contained in the research model (Figure 4.1) this is shown in Table 6.2.

Research Model Variables	INTERPROD Publications Variables
R&D/Marketing Integration	R&D/marketing integration, Souder et al., 1997
Project Management Skills	Project management, Souder et al., 1997
Top Management Support	Top management support, Wren et al., 2000
Development Process Proficiency	Technical Proficiency, Souder et al., 1997
New Product Characteristics	Entry Strategies, Yap & Souder, 1994 Performance superiority, Souder & Song, 1997 Technical superiority, Souder & Song, 1997 Radicalness, Souder & Song, 1997 Compatibility, Souder & Song, 1997
Market Characteristics	Users eagerness to adopt, Souder & Song, 1997 Market growth potential, Souder & Song, 1997 Intensity of competition, Souder & Song, 1997 Customer orientation, Wren et al, 2000
Marketing and Technological Skills	Technical skills, Souder et al., 1997 Technical Proficiency, Souder et al., 1997
Marketing and Technological Synergy	Marketing skills, Souder et al., 1997
Methods of Organising NPD	Organisation for NPD, Lee et al., 2000 Organisational processes, Lee et al., 2000

Table 6.2 Research Model vs. INTERPROD Publication Variables

The first three publications included in this chapter (Yap & Souder, 1994; Souder & Song, 1997; Souder et al. 1997) examined data from small US firms. These small firms were defined as having fewer than 500 employees. Within the Irish data set small firms are defined as having fewer than 100 employees. In fact, there are only three Irish firms in the study with more than 500 employees. Therefore, in comparing results from the analysis of the Irish data with results from these three papers it is more suitable to compare small US firms with the complete Irish sample rather than with the small Irish firms.

6.2 INTERPROD Results

In this section the Irish INTERPROD data is analysed to determine the impact of the 18 factors listed in Table 6.1 on new product success in Irish firms. These 18 factors were extracted from five INTERPROD publications and so this section will be divided into five subsections, one for each INTERPROD publication.

Many of the publications described in Chapter 3 examined the impact that market and technical uncertainty had on NPD management. Market and technical uncertainty was not considered when the Irish INTERPROD data was collected; therefore, the Irish sample is not balanced in terms of market or technical certainty as is shown in Table 6.3.

	Market Uncertainty		**Technical Uncertainty**	
	Low	**High**	**Low**	**High**
All Firms (n = 63)	50	13	43	20
Small Firms (n = 37)	29	8	26	11
Large Firms (n = 26)	21	5	17	9

Table 6.3 Market conditions in the Irish data

In many of the INTERPROD publications, successful projects were defined as those that exceeded the expectations of their management teams. Table 6.4 shows the breakdown of the Irish data in terms of project success. In the analysis that follows, the 9 projects that were consistent with expectations will be omitted. The 29 projects that scored 1 or 2 will be considered to be failures and the 25 that scored 4 or 5 will be considered successful; the total number of projects in Irish firms analysed will be 54.

Score	Commercial outcome of the product	Number of small firm projects	Number of large firm projects	Total Projects
1	Far below our expectations	15	11	26
2	Slightly below our expectations	2	1	3
3	Consistent with our expectations	7	2	9
4	Slightly above our expectations	7	3	10
5	Far above our expectations	6	9	15

Table 6.4 Success of Irish projects

6.2.1 Entry Strategies: Yap & Souder, 1994

Yap & Souder (1994) analysed small high-technology electronics firms under conditions of high and low technical and market uncertainty (see Table 3.5 in Chapter 3). The factors from their paper that are examined here are Entry Strategies. The analysis will be carried out on the complete Irish data set. The number of projects from Irish firms that fit into each of the technical and market uncertainty groups is summarised in Table 6.5.

	Market Uncertainty		Technical Uncertainty	
	Low	High	Low	High
All Firm Projects (n = 54)	42	12	38	16

Table 6.5 Sample sizes for technology sources and entry strategies

Table 6.6 shows the relationships between entry strategies and product success. These results do not agree with those published by Yap & Souder. Positioning products based on 'product class dissociation' is positively linked with success in large Irish firms; the relationship was negative in small US firms. Also, 'emphasising solutions and benefits' is linked with success in Irish firms, but not in small US firms.

	Correlation Coefficients between factor scores and project outcomes^{a}			
	Market Uncertainty		Technical Uncertainty	
	Low	High	Low	High
Entry Strategies – All Projects:				
Positioning: Against another product [C28]	-.140	.000	-.040	-.187
Positioning: Product class dissociation [C29]	.352^{c}	-.262	·258^{e}	.208
Distribution: Developer's own salesforce [C30]	-.022	.119	-.045	.200
Distribution: Manufacturer's representatives	.115	-	-.043	-.058
[C32]	.314^{c}	.398^{d}	.154	.568^{c}
Positioning: Benefits and solutions emphasised		-.031		
[C25]				
Entry Strategies – Small Projects:				
Positioning: Against another product [C28]	-.261	-.255	-.260	-.213
Positioning: Product class dissociation [C29]	.215	-.094	.113	.429
Distribution: Developer's own salesforce [C30]	-.074	.087	.021	-.173
Distribution: Manufacturer's representatives	.008	-.283	-.019	-.238
[C32]	.295^{e}	.177	.324^{e}	.429
Positioning: Benefits and solutions emphasised				
[C25]				
Entry Strategies – Large Projects:				
Positioning: Against another product [C28]	-.048	.236	.176	-.207
Positioning: Product class dissociation [C29]	.497^{c}	-.707	.441^{d}	-.007
Distribution: Developer's own salesforce [C30]	.024	.236	-.112	.387
Distribution: Manufacturer's representatives	.234	-.577	-.122	.312
[C32]	.389^{d}	-.577	.076	.779^{c}
Positioning: Benefits and solutions emphasised				
[C25]				

Correlation coefficients statistically significant at: $^{a}p \leq .005$; $^{b}p \leq .01$; $^{c}p \leq .05$; $^{d}p \leq .10$; $^{e}p \leq .20$;

Table 6.6 Entry Strategies

6.2.2 Design and Market Parameters: Souder & Song, 1997

Seven factors, divided into two categories, i.e. design parameters and market parameters were analysed by Song & Souder (see Table 3.7 for a definition of the 7 factors). This analysis is repeated on the complete Irish data set in Table 6.7.

	Spearman Correlation Coefficients					
	All Firms		**Small Firms**		**Large Firms**	
	Market Uncertainty		**Market Uncertainty**		**Market Uncertainty**	
Parameter	**Low**	**High**	**Low**	**High**	**Low**	**High**
Design						
Parameters:	·394[b]	-.344	.370	-.085	.522[c]	-.894
Performance	.241	-.426	.090	-.181	.357	-.707
Superiority	.252	-.561	-.119	-.334	.580[b]	-.943
Technical	.054	-.170	.169	-.167	-.078	-.943
Superiority						
Radicalness						
Compatibility						
Market						
Parameters	-.015	.339	-.004.	-.167	-.055	1.00[a]
Intensity of	.188	.085	-.142	.167	.507[d]	.000
Competition	193	.373	.336	.428	.084	.894
Market Growth						
Potential						
Users Eagerness to						
Adopt						

Statistically significant at: $^a p \leq .001$; $^b p \leq .01$; $^c p \leq .02$; $^d p \leq .05$; Two-tailed.

Table 6.7 Analysis based on Song & Souder, 1997

The impact of *design parameters* on new product success in the Irish data was not the same as that reported by Souder & Song (see Table 3.8 in Chapter 2). Table 6.7 shows that under conditions of low market uncertainty 'performance superiority' correlates with commercial success, particularly in large firms. While this result is intuitively sound, it was only found by Souder & Song under conditions of high market uncertainty.

By contrast, there was a negative correlation between 'performance superiority' and commercial success under conditions of high technical or market uncertainty for all Irish firms. Table 6.7 shows that all design parameters are negatively correlated with commercial success under conditions of high uncertainty – this is not consistent with the findings of Souder & Song. 'Radicalness' was negatively correlated with success under conditions of low market uncertainty in the firms studied by Souder & Song – this was also found in small Irish firms but the opposite was seen in the large firms in Table 6.7.

The results relating to *market parameters* in Table 6.7 are somewhat closer to those found by Souder & Song. The two parameters that correlate with commercial success in large firms in the table above, 'market growth potential' and 'intensity of competition', were also found to correlate with success in the large US and Japanese firms examined by Souder & Song.

6.2.3 Technical and Market Proficiency: Souder, Buisson & Garrett, 1997

Souder et al. defined 6 component items and measured the correlation between these items and product performance. Table 6.8 replicates the analysis carried out by Souder et al. (see Table 3.13 in Chapter 3) on the Irish data.

	Spearman Correlation Coefficients and Mean Scores					
	All Firms (N=54)		Small Firms (N=30)		Large Firms (N =24)	
Parameter	**Coefficient**	**Mean**	**Coefficient**	**Mean**	**Coefficient**	**Mean**
Project Manager Characteristics	.158	3.75	.114	3.76	.163	3.74
Technical Skills	.233	3.86	.083	3.86	.401	3.87
NPD Marketing Proficiency	.507[a]	2.91	.610[a]	2.86	.381	2.99
R&D/Marketing Integration	.439[a]	3.46	.465[b]	3.45	.380	3.49
Marketing Skills	.594[a]	3.00	.667[a]	2.95	.497[c]	3.06
NPD Technical Proficiency	.526[a]	3.50	.613[a]	3.46	.420[c]	3.55

[a] Significant at $p \leq .001$: [b] Significant at $p \leq .01$: [c] Significant at $p \leq .05$

Table 6.8 Analysis based on Souder, Buisson & Garrett, 1997

T-tests were also carried out between the mean value of each parameter in small and large firms and no significant differences were found. The results above are similar to those reported by Souder et al. who found that all the parameters they tested, with the exception of 'technical skills', correlate with product performance. The one difference is the lack of significant 'project manager characteristics' in the Irish data.

6.2.4 NPD Organisational Characteristics: Lee, Lee & Souder, 2000

Lee et al. carried out a subgroup analysis of the methods of organising for NPD (see section 3.3.11 in Chapter 3). This involved splitting the sample based on whether or not a particular method of organisation was used and calculating the mean level of success for each group. If there is a significant difference between the mean success of firms who used or did not use a particular organisational method it is assumed that the method is of significance. The results of this subgroup analysis are shown in Table 6.9. Table 6.9 also shows that with the exception of the use of a 'dyad' (a strong interpersonal alliance between two persons of different functions), no method of organising for NPD had a significant impact on the level of success of the project. In the Irish sample, dyads were found to be effective but the sample size is too small to draw any firm conclusions.

All Firms (N = 54)	**Used**	**Non-Used**	**t-value**
NPD Department	3.00 (N= 21)	2.55 (N= 33)	.900
NPD Manager	2.86 (N= 22)	2.63 (N= 32)	.474
Venture Team	2.75 (N= 12)	2.71 (N= 42)	.060
Task Force	3.00 (N= 8)	2.67 (N= 46)	.468
NPD Committee	3.60 (N= 5)	2.63 (N= 49)	1.144
Matrix Organisation	2.63 (N= 8)	2.74 (N= 46)	-.163
Top Management Control	2.69 (N= 16)	2.74 (N= 38)	-.091
Dyad	4.67 (N= 3)	2.61 (N= 51)	1.971[a]
Small Firms (N = 30)			
NPD Department	2.67 (N= 9)	2.52 (N= 21)	.203
NPD Manager	2.70 (N= 10)	2.50 (N= 20)	.293
Venture Team	2.70 (N= 10)	2.50 (N= 20)	.293
Task Force	2.75 (N= 4)	2.54 (N= 26)	.223
NPD Committee	(N= 0)	2.57 (N= 30)	-
Matrix Organisation	3.00 (N= 2)	2.54 (N= 28)	.360
Top Management Control	2.67 (N= 12)	2.50 (N= 18)	.253
Dyad	5.00 (N= 1)	2.48 (N= 29)	1.453

Large Firms (N = 24)			
NPD Department	3.25 (N= 12)	2.58 (N= 12)	.850
NPD Manager	3.00 (N= 12)	2.83 (N= 12)	.209
Venture Team	3.00 (N= 2)	2.91 (N= 22)	.063
Task Force	3.25 (N= 4)	2.85 (N= 20)	.375
NPD Committee	3.60 (N= 5)	2.74 (N= 19)	.896
Matrix Organisation	2.50 (N= 6)	3.06 (N= 18)	-.609
Top Management Control	2.75 (N= 4)	2.95 (N= 20)	-.187
Dyad	4.50 (N= 2)	2.77 (N= 22)	1.239

Statistically significant at: [a] $p < 0.1$

Table 6.9 Impact of methods for organising for NPD on NPD performance. [B17]

Lee, Lee & Souder (2000) also examined a range of organisational process factors and their impact in NPD performance. This analysis is repeated in the Irish data in Table 6.10.

	Spearman Correlation Coefficients and Mean Scores					
	All Firms (N=54)		Small Firms (N=30)		Large Firms (N =24)	
Parameter	**Coefficient**	**Mean**	**Coefficient**	**Mean**	**Coefficient**	**Mean**
Level of procedure formality [B13]	.074	3.56	.043	3.40	.038	4.04
Level of authority concentration [B14]	.064	4.29	-.023	4.27	.115	4.29
Level of participation in decisions [B15]	.205	4.05	.095	4.10	.302	4.08
Degree of organicity [B16]	.060	3.59	.008	3.80	.167	3.33
R&D/Marketing Integration [B18 – B22]	.430[a]	3.27	.460[c]	3.32	.413[c]	3.20

[a] Significant at $p \leq .001$: [b] Significant at $p \leq .01$: [c] Significant at $p \leq .05$

Table 6.10 Organisational Processes

Only 'R&D/Marketing integration' was linked with success in the Irish data presented in Table 6.10.

6.2.5 Market Orientation: Wren, Souder &, Berkowitz, 2000

Wren et al.(2000) examined the impact of several composite variables on success and also the impact of three variables (project management skills, top management support and integration) on market intelligence and customer orientation. Their results are shown in Table 3.21, in Chapter 3. Table 6.11 repeats this analysis concentrating on three variables, i.e. market intelligence, customer orientation and top management support. (Note: these three variables were selected to avoid overlap with other variables analysed in this chapter.)

	All Firms (N=54)	**Small Firms (N=30)**	**Large Firms (N =24)**
Dependent Variable = Success			
Market Intelligence [C50, E6]	.535[a]	.489[a]	.519[a]
Customer Orientation [C6, C23]	.509[a]	.692[a]	.290
Dependent Variable = Market Intelligence			
Top Management Support [B23, E8]	.411[a]	.192	.576[a]
Dependent Variable = Customer Orientation			
Top Management Support [B23, E8]	.118	.196	.029

[a] Significant at $p \leq .01$

Table 6.11 Market Intelligence, Customer Orientation and Top Mgmt. Support

The results obtained for market intelligence and customer orientation agree with those reported by Wren et al., with the exception of the lack of correlation between customer orientation and success in large Irish firms. A definite relationship was found between top management support and market intelligence in large but not small firms, while no relationship was found between top management support and customer orientation. In this respect the large Irish firms were similar to the US firms explored in the study by Wren et al. This is not surprising as many of the large Irish firms were Irish-managed subsidiaries of US multinationals.

6.3 INTERPROD Discussion

Within this section, results from each of the five papers included in this chapter will be discussed and common themes will be summarised.

6.3.1 Entry Strategies: Yap & Souder, 1994

Differences were seen between the entry strategies that are linked with success in US and Irish firms. 'Product class dissociation' is a positive positioning strategy for the Irish firms but not the US firms. This could be due to the fact that the smaller Irish firms frequently operate in market niches where they succeed by offering differentiated products. This explanation is supported by the positive link between success and 'emphasising the benefits and solutions' in Irish firms.

The negative effect of using manufacturers' representatives to distribute new products in cases of high-market uncertainty is possibly related to the isolated nature of Irish firms. Using a manufacturers' representative can be a useful strategy for distributing new products if sufficient support can be given to the representative. However, in the Irish situation where the vast majority of new products are exported to remote markets, it is possible that using a representative who may not fully understand a new product is less likely to be a successful strategy. This suggests that the location of a particular firm can have an impact on its success strategies.

6.3.2 Design and Market Parameters: Souder & Song, 1997

At a high level, the findings of Souder & Song are supported in the Irish data – i.e. firms should adopt different market and product strategies based on the level of market uncertainty they experience. However, the nature of these differences is not the same in Irish and US/Japanese firms. Under conditions of high market uncertainty none of the design parameters examined by Souder & Song were linked with success in Irish firms; within the US/Japanese data several significant links were found. However, to explain this finding, the data must be examined in more detail.

Only 13 of the 63 Irish projects studied were launched under conditions of high-market uncertainty. Of these 13 projects, the commercial outcomes of 9 were rated to be far below expectations; 1 met expectations; and 3 were far above expectations. The results shown in Table 6.6 demonstrated that the 3 successful projects launched into uncertain markets were not superior in terms of performance or technical content; nor were they more radical. Further examination of the data shows that Irish firms launched only 7 products that used uncertain technology for uncertain markets and of these products 5 were judged to perform far below commercial expectations, see Figure 6.1

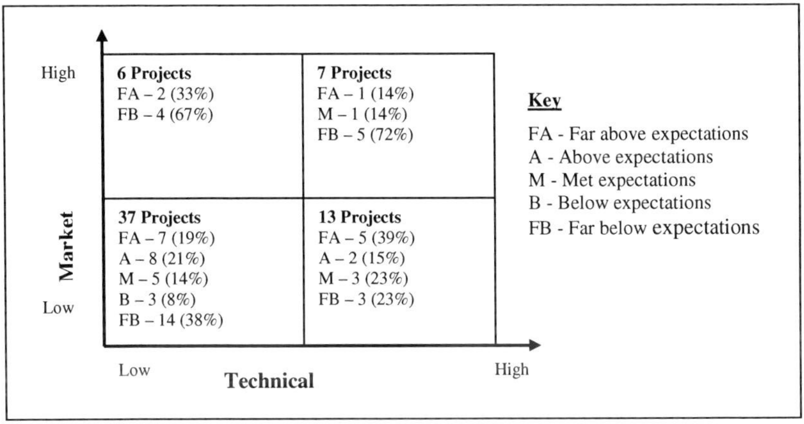

Figure 6.1 Uncertainty vs. Success in Irish Firms

As mentioned at the start of this chapter, the Irish data was not balanced in terms of market or technical uncertainty. The benefit of this is that it allows some comment to be made about the type of new product that Irish electronics firms develop. Figure 6.1 shows that the majority of new products (59%) developed by Irish firms were launched under conditions of low technical capability and market uncertainty. These could be classified as regular or incremental product innovations (Clark, 1983; Pearson, 1991).

Figure 6.1 shows that Irish firms are reluctant to enter new markets and perhaps, justifiably so; the success rate for new products launched under conditions of high market uncertainty (31%) is half of that for products launched under conditions of low market uncertainty (60%). It is also worth noting that the area in which Irish firms have most success (77%) is when they develop products under conditions of low market uncertainty and *high* technical uncertainty. In summary, Irish firms are good at developing products based on new technologies but not at developing products for new markets.

6.3.3 Technical and Market Proficiency: Souder, Buisson & Garrett, 1997

There is considerable agreement between the results of Souder et al. and the results presented above as based on the Irish data. Four of the five factors that correlated significantly with success in the US and NZ firms were also significant for the Irish firms. The one variable that was significant in the US and NZ but not in Ireland was 'project manager characteristics'. Despite the fact that the Irish firms reported similar levels of Project Management Skills to the other two countries, project success was not linked with increased Project Management Skills in Irish firms. A possible explanation for this is that Irish firms are not fully embracing the role of the project manager and thus not benefiting from increased Project Management Skills.

Technical skills were not linked with new product success in the US, NZ or in Ireland. Souder et al. explain this by saying that all the projects they examined had relatively high technical skill scores. The previous discussion shows that Irish firms are also good at developing technically difficult projects. Technical proficiency would appear to be a prerequisite for high technology firms rather than a definite success factor.

It is also interesting to note that the mean scores of the Irish firms for each of the six items measured are closer to the scores reported by the US firms than they are to those reported by the NZ firms. This shows clearly the influence of the many US electronics multinational firms located in Ireland on the management of NPD within Irish firms.

The results presented in Table 6.8 show two factors that were linked with success in small but not large firms. These were 'NPD marketing proficiency' and 'R&D/Marketing Integration'. In both cases the mean scores reported by the large firms were higher than those reported by the small firms therefore it is possible that these factors are consistently well-managed in large firms and are therefore not differentiators between new product success and failure.

The final two variables, 'Marketing Skills' and 'NPD Technical Proficiency' were linked with success in both small and large firms; though, for both variables, the correlation coefficients were more significant for small firms. The mean scores for these variables were higher in large firms than in small firms. Again, this suggests that 'Marketing Skills' and 'NPD

Technical Proficiency' are less likely to be determinants of success in large firms than in small firms.

6.3.4 NPD Organisational Characteristics: Lee, Lee & Souder, 2000

Table 6.8 shows that the most commonly used method of organising for NPD in Irish firms was under the supervision of a NPD manager (41%). This was also a popular method in the US and Korean firms studied. By contrast, *small* Irish firms favoured the use of top management control (40%), a method used by 30.2% of the US firms in the study. Large Irish firms were most likely to use NPD departments (50%) and NPD managers (50%); these were also the most popular methods used by Korean firms.

Lee et al. found significant differences between Korea and the US in the level of authority concentration and R&D/Marketing integration. Additionally, they found that 1) level of participation in decisions, 2) degree of organicity and 3) R&D/Marketing integration correlated with success in Korean firms while 1) level of authority concentration, 2) level of participation in decisions and 3) R&D/Marketing integration correlated with success in the US.

Repeating the analysis of Lee et al. on the Irish data yielded few unexpected results. The most popular method of organising NPD projects in small firms was through top management control, while in large firms NPD departments and managers were most frequently used. These findings are supported by previous publications (Santarelli & Sterlacchini, 1990; Roper, 1996). Also, none of the organisational methods used were linked with new product success in Irish firms.

Only one of the organisational process variables was linked with success in Irish firms; this was the level of R&D/Marketing integration and was found to be significant for both small and large Irish firms. (In Section 6.2.3 the analysis replicating the publication by Souder et al. also linked R&D/Marketing integration with success although a slightly different measure was used.) Lee at al. also found a link between R&D/Marketing integration and success in both Korea and the US, as have several other authors (Griffin & Hauser, 1996; Song & Parry, 1997; Leendres & Weirenga, 2002).

The lack of any link between the other organisational processes examined is of interest. It appears that, with the exception of the R&D/Marketing interface, the way in which Irish firms organise for NPD projects has little impact on the success of these projects. The factors that have been linked with success include several product parameters, technical and marketing proficiency and marketing skills. This data suggests that possessing the relevant skills and proficiencies for NPD is more important for Irish firms than how these skills are organised.

6.3.5 Market Orientation: Wren, Souder &, Berkowitz, 2000

Repeating the analysis of Wren et al. on the Irish data demonstrates once again that the management of NPD in Irish electronics firms is influenced by the presence of many US electronics multinationals. Market intelligence and customer orientation were linked with success in Irish firms; Wren et al. also found this to be the case in the countries they studied. One exception to this was the finding that customer orientation was not a significant factor for large Irish firms. There could be several explanations for this. Many of the large Irish firms are subsidiaries of US firms; it is possible that customer information is supplied to the Irish firms through their US parent and therefore customer orientation is not important for the Irish firms. It is also possible that the nature of the NPD carried out by large Irish firms is such that being close to their customers is not as critical for them as it would be for most firms. For example, large Irish firms are mainly involved in following market trends – i.e. they are fast followers as opposed to market leaders.

Top management support was not linked with either market intelligence or customer orientation in small Irish firms. Wren et al. found similar results in Korean and Belgian firms and suggested that could be due to the culture in these countries, one where individual behaviour is determined more by social rules than management direction. The explanation is probably different in the small Irish firms. Previous data shows that many NPD projects in small Irish firms come under direct top management control, see Section 6.2.4. Therefore, it is likely that there is a very high level of top management support for all NPD activities in small firms; in other words, top management support is not a differentiating factor.
In large Irish firms top management support is linked with market intelligence, but not with customer orientation. Wren at al. reported similar findings for US firms; this again suggests a strong US influence in large Irish firms but also shows that even the top management in these

firms do not promote customer orientation. In other words, the lack of customer orientation in large Irish firms is not simply an oversight, but comes from the highest level in the firm.

6.4 Summary of Preliminary INTERPROD Analysis

The 18 factors listed in Table 6.1 have been analysed to determine their impact on new product success in Irish firms. A summary of this analysis is shown in Table 6.12. The relationship between the factors examined in this chapter and the variables in the research model is also shown in Table 6.12.

Four key findings have emerged from this analysis.

1. *There are few NPD success strategies that apply in all situations – the successful management of NPD is highly context dependent.* There is remarkably little consensus in the 13 papers published about the findings of the INTERPROD study as to how NPD should be managed. The analysis of the Irish data produced another set of results about which NPD strategies and management practices are linked with success in small and large Irish firms. The overall framework for the INTERPROD study - presented in Chapter 3- proposed that successful NPD management practices are dependent on the level of technical and market uncertainty. The analysis in this chapter shows that firm location (country) and size also have an impact on which management practices are linked with success.
2. *Irish firms are involved mainly in 'regular' or incremental NPD projects.* The breakdown of the 63 NPD projects undertaken by firms involved in this study shows that almost 60% of new products used existing technology to address existing markets. This data was collected during a period of unprecedented economic growth in Ireland when it might have been expected that high-technology firms would be developing new markets; however, this does not appear to have been the case. High levels of product innovation are associated with increased competitive advantage (Tidd et al, 1997). The incremental nature of product innovation in the firms included in this study suggests that Irish electronics firms are not protecting their competitive positions.

Factor	Comment	Corresponding Model Variable
1. Entry Strategies: Source: Yap & Souder, 1994	The entry strategies found to be successful in the US firms did not apply in the Irish firms; this is probably due to the high levels of export in Irish firms and also to their smaller size. Irish firms launch few products into uncertain markets and have limited success with them. None of the design parameters examined was linked with success under conditions of high market uncertainty. Under conditions of low market uncertainty performance and technical superiority and radicalness were linked with success.	**New Product Characteristics**
2. Performance superiority: Source: Souder & Song, 1997		
3. Technical superiority: Source: Souder & Song, 1997		
4. Radicalness: Source: Souder & Song, 1997		
5. Compatibility: Source: Souder & Song, 1997		
6. Users eagerness to adopt: Source: Souder & Song, 1997	Selecting customers who are eager to adopt is the strategy that most strongly linked with success in both high and low market uncertainty.	**Market Characteristics**
7. Market growth potential: Source: Souder & Song, 1997		
8. Intensity of competition: Source: Souder & Song, 1997		
9. Market proficiency: Source: Souder & Buisson & Garret, 1997	This is significantly correlated with success in small Irish firms but not large Irish firms, though large firms report higher levels of market proficiency.	**Marketing and Technological Skills**
10. Technical Proficiency: Source: Souder & Buisson & Garret, 1997	Significantly correlates with NPD success in Irish US and NZ firms.	

11. Technical skills: Source: Souder & Buisson & Garret, 1997	This is not a differentiator between success and failure, a finding consistent with Souder et al.	
12. Marketing skills: Source: Souder & Buisson & Garret, 1997	Significantly correlates with NPD success in Irish US and NZ firms.	
13. Project management: Source: Souder & Buisson & Garret, 1997	Not linked with success in Irish firms suggesting that the role of project manager is not sufficiently utilised in Irish firms.	**Project Management Skills**
14. R&D/marketing integration: Source: Souder & Buisson & Garret, 1997	This is significantly correlated with success in small Irish firms but not large Irish firms, though large firms report higher levels of market proficiency.	**R&D/Marketin g Integration**
15. Organisation for NPD: Source: Lee, Lee & Souder, 2000	The method for organising for NPD was not linked with NPD success in Irish firms.	**Methods of Organising NPD**
16. Organisational processes: Source: Lee, Lee & Souder, 2000	R&D/Marketing integration is the only organisational process that is linked with NPD success.	
17. Customer orientation: Source: Wren, Souder & Berkowitz 2000	Market intelligence is linked with success in large and small Irish firms. Customer orientation is linked with success in small firms but not in large firms.	**Market Characteristics**
18. Top management support: Source: Wren, Souder & Berkowitz 2000	Top management support is not a differentiating factor in small Irish firms but it is linked with increased market intelligence in large Irish firms.	**Top Management Support**

Table 6.12 Summary of INTERPROD Analysis (continued)

3. *The management of NPD in Irish electronics firms is influenced by a strong US presence within the electronics sector in Ireland.* Many of the large firms included in this study are subsidiaries of US multinationals and former employees of US firms founded

many of the small firms. Therefore it is not surprising that there is a US influence seen in how NPD is managed in Irish electronics firms. This influence has probably been positive increasing the skill level of Irish R&D managers. But the INTERPRDO project has shown that the successful management of NPD is dependent on firm location therefore it may not always be appropriate to assume that NPD management practices that work in the US will work in Ireland.

4. *Several factors, including technology sources, organisational factors, market parameters, and technical skills were not linked with new product success in Irish firms.* Many of these factors were linked with success in other countries but not in Ireland. Again, this emphasises the contextual nature of NPD success factors but it also prompts a question about which NPD management practices are linked with success in the Irish firms and particularly in small Irish firms.

6.4.1 Updated Research Model

The findings summarised above are reflected in an update model of NPD management, Figure 6.2. The updated model can now detail the impact of the NPD context, firm size and location, on NPD management and success. For clarity, technical and market uncertainty have been omitted from the updated model, though both of these variables have been found to impact the management of NPD they are not central to this study.

The updated model presented in Figure 6.2 builds on the proposed model presented in Figure 4.1. Only the project level variables have been examined. The updated model shows that the management of some variables and their effect on new product success is affected by firm size while that of other variables is affected by firm location.

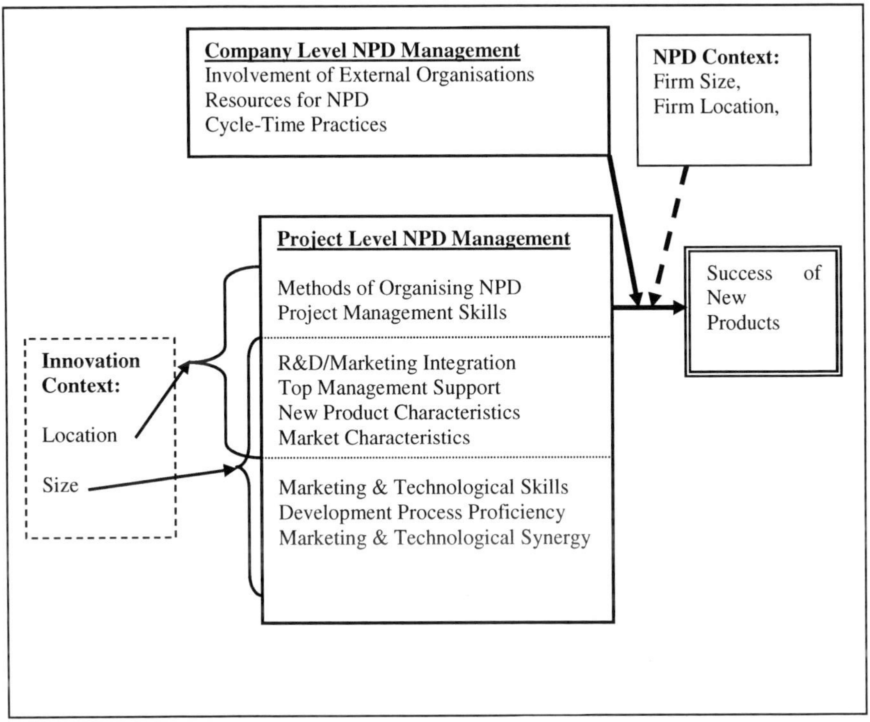

Figure 6.2 Updated model for NPD management based on INTERPROD

The updated model proposes the relationships between the proposed nine project level determinants of success and firm size and location. These relationships are summarised in Table 6.13.

This preliminary analysis of the Irish INTERPROD data has served to position the management of NPD in Ireland within an international context. It has also reinforced one of the fundamental elements of the research model – 'good practice' in managing NPD is dependant on the context within which firms operate. The analysis in this chapter did not set out to 'test' the research model; none of the company level management practices were examined. Figure 6.2 shows which context variables have an impact on which management practices; this is based on the information in Table 6.12.

Project Level Variable	Impact of Firm Size and Location
Development Process Proficiency:	In Ireland Development Process Proficiency is more closely linked with success in small firms than in large firms.
Marketing and Technological Synergy:	In Ireland Marketing and Technological Synergy is more closely linked with success in small firms than in large firms.
Methods of Organising NPD:	Unlike in Korea and the US, there is no link between the method of organising for NPD and success in Irish firms
Project Management Skills:	Unlike in New Zealand and the US, there is no link between project management skills and success in Irish firms.
R&D/Marketing Integration:	In Ireland R&D/Marketing integration was only linked with success in small firms.
Top Management Support:	In Ireland top management support is linked with increased market intelligence in large firms.
New Product Characteristics:	Few new products characteristics were linked with success in Irish firms and relationships only existed under conditions of low market uncertainty and depend on firm size.
Market Characteristics:	Several market characteristics were linked with success but the relationships were different for large and small firms.
Marketing and Technological Skills:	Different skills were linked with success in small and large firms.

Table 6.13 Project Level Variables vs. Firm Size and Location

The analysis in this chapter provides some insights into the management of NPD in small Irish electronics firms, however it has been limited. Company level management practices have not been addressed and many of the project level practices have not been explored in sufficient detail. The analysis and discussion presented here is insufficient to fully answer the research question therefore it is necessary to explore the Irish INTERPROD data further. In chapters 7 and 8 the Irish INTERPROD data is examined to explore the management of the variables defined in the research model and the impact that they have on new product success in small Irish electronics firms.

6.5 Conclusion

This chapter presented a ‘first cut’ of the Irish INTERPROD data based on the analysis undertaken in 5 previously published INTERPROD papers that were considered to be relevant to the Irish situation. The results show that NPD management and success factors in Irish firms are not the same as those in firms in other countries and also that management and success factors are different in small and large firms.

The next stage in this study is to explore the extent of the differences highlighted here. As discussed in Chapter 5 this could be done by comparing Ireland with countries that are known to be very different but it is considered more useful (and practical) to compare Ireland with a country that would be expected to be similar – the UK.

Therefore, Chapter 7 will explore the INTERPROD data in much more detail, presenting firm and project level NPD management practices and determining which practices are linked to success in small and large firms in Ireland and the UK.

CHAPTER SEVEN

INTERPROD RESULTS IRISH & UK DATA

7.1 Introduction

This chapter reports the results of the INTERPROD research in Ireland and compares it to data collected from electronics firms in the UK. As described in Chapter 5 the INTERPROD research tool has two types of questionnaire - those collecting data about the overall management of product development within the firm and those collecting data about how specific NPD projects were managed. The chapter is divided into two main parts;

- *Section 7.2* reports on the company-level findings,
- *Section 7.3* reports on the project specific results,

Data from two groups of firms, Irish electronics firms and UK electronics firms, are analysed and discussed. Each group of firms is further divided into small firms and large firms. For the purpose of this study small firms are defined as those with fewer than 100 employees. Table 7.1 summarises the number of firms and projects within each group of interest. Note: successful projects have been defined in Chapter 5 as projects that met or exceeded the expectations of the firm.

The results presented in Section 7.2 are mainly in the form of median scores for the six groups of firms of interest. These medians are compared and statistically significant results are noted. The data presented in Section 7.3 show the degree to which the management practices addressed in the INTERPROD questionnaires correlate with new product success in each of the six groups of firms.

	Number of Firms	Number of Projects	Successful Projects	Unsuccessful Projects
All Irish Electronics Firms	36	63	34	29
Small Irish Electronics Firms	22	37	20	17
Large Irish Electronics Firms	14	26	14	12
All UK Electronics Firms	25	39	27	12
Small UK Electronics Firms	5	10	6	4
Large UK Electronics Firms	20	29	21	8

Table 7.1 Sample of Firms in the INTERPROD Study

For ease of comparison with previous INTERPROD research, and to help the reader to place these results in context, question codes are provided in square brackets next to each parameter analysed in this chapter. These question codes can be used to locate a question within the seven INTERPROD questionnaires in Appendix 1. Additionally, where data has been presented in graph form the corresponding 'raw' data is given in Appendix 5. A more detailed description of the layout of this chapter and the analysis of the data is given below.

7.1.1 Data Analysis

Most of the data presented in this chapter are ordinal - the majority of the INTERPROD questions required respondents to chose between five different answers and results were recorded as the number of the response chosen. Where ratio data have been collected, e.g. number of employees, normality tests (Shapiro-Wilkes) have been applied, the results of which show that even these data are not normally distributed. Therefore, throughout this chapter non-parametric statistical tests are used. Results are presented as medians (with ranges given in Appendix 5) rather than as means with standard deviations.

The objective of the data analysis in this chapter is to provide answers to the research questions posed in Chapter 5:

How do small Irish electronics firms manage the development of new products, and are the success factors that apply to small Irish firms different from those that apply to large Irish firms?

1.a Are the differences between small and large Irish electronics firms also seen between small and large UK electronics firms?

1.b How does NPD management in the Irish electronics sector, as measured by the INTERPROD research tool, compare with that in the UK?

The analysis in this chapter consists of comparing the responses of small and large, UK and Irish electronics firms to the INTERPROD questionnaires. Figure 7.1 shows how the different groups of firms described in Table 7.1 are compared in Sections 7.2 and 7.3. The three tests shown in Figure 7.1 compare the following groups of firms:

Test 1: Small Irish Electronics vs. Large Irish Electronics firms
Test 2: Small UK Electronics vs. Large UK Electronics firms
Test 3: Irish Electronics vs. UK Electronics firms

Test 1 addresses the primary research question – Question 1. Test 2 addresses the subsidiary, research question - Question 1.a. Test 3 is included to further inform the discussion on the degree to which findings can be generalised beyond the Irish setting – Question 1.b.

Section 7.2 makes extensive use of the Mann-Whitney *U* test to analyse company-level data. The Mann-Whitney *U* test is a non-parametric alternative to the *t* test and is used to compare two sets of data. Therefore in Section 7.2, Tests 1, 2 and 3 illustrated in Figure 7.1, will be implemented primarily through the use of the Mann Whitney *U* test.

Section 7.3 presents project-level data, in this case data from successful and unsuccessful projects have been collected. Rather than simply comparing the responses of each of the six groups of firms of interest, responses are correlated with the commercial success level of each project to determine which management practices correlate with success within each group of firms. Again a non-parametric test, this time Spearman's rank order coefficient, is used. In most of Section 7.3, Tests 1, 2 and 3 will be addressed in the 'Observations' sub-sections as opposed to being explicitly presented in tabular format.

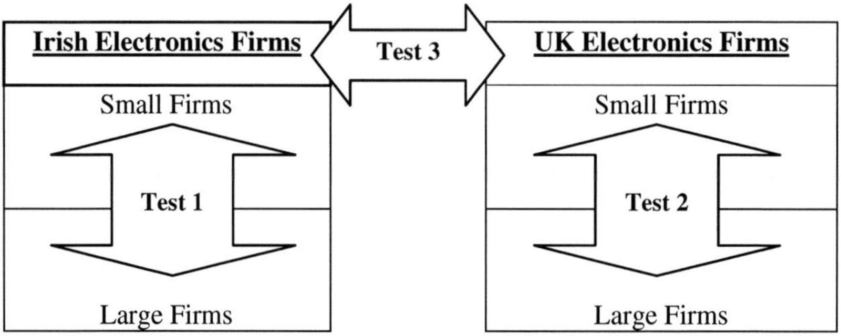

Figure 7.1 Data Comparisons in Sections 7.2 and 7.3

As discussed above, the most statistically robust way to analyse the data collected from the different groups of firms is to compare the medians of each set of data. However there are occasions when presenting only medians does not provide sufficient insight into the responses of each group of firms. This is especially true when the data being presented are based on a Likert scale. On these occasions, means will also be included. A justification for presenting the mean of these data follows.

The median of a set of data is the value for which 50% of values are greater and 50% are less. Many of the INTERPROD questions required the respondent to select an answer from one of five options; responses were then recorded as a number between 1 and 5. For many questions the majority of responses were in the mid region, i.e. 3 or 4, and the median of a set of such responses is either 3 or 4. This analysis is not particularly useful in examining differences and trends between groups. For example, one section of the INTERPROD questionnaire asks respondents to indicate their degree of concern for a list of potential challenges. These data are examined by determining which challenges are of most concern to each group of firms. When medians are used in this analysis all the challenges that are of serious concern to respondents have a median of 4.00, allowing limited differentiation between groups. However, if the mean scores are calculated for each group it is potentially easier to differentiate among the key challenges for the different groups and possibly identify interesting directions for future research.

7.1.2 Chapter Layout

The first part of this chapter, *Section 7.2*, presents the *company*-level results of the INTERPROD questionnaires (INT-A, INT-F and INT-L) that were administered to electronics firms in the UK and Ireland. The nature of these questionnaires and the manner in which the data were collected is described in Chapter 5. Section 7.2 is further divided into six areas (sub-sections):

1. NPD Context – INT-A, INT-L
2. Performance Measures – INT-A, INT-L
3. Development Process Formality – INT-L
4. Involvement of External Organisations – INT-A
5. Resources for NPD – INT-A
6. Cycle-time Practices – INT-F

Within each of these sub-sections median scores are presented, in a tabular format, for each of the six groups of firms of interest as listed in Table 7.1. For clarity of presentation the ranges of the INTERPROD data are not included in the tables in this chapter, but they can be found in Appendix 5. A second table is also included in each sub-section comparing the median scores as specified in Tests 1, 2 and 3. Finally, each subsection contains a list of general observations on the data presented. Chapter 8 contains a full discussion of these observations. In summary, each sub-section in Section 7.2 contains the following:

- A brief description of the data about to be presented.
- Brief observations on the data presented.
- A table containing median scores.
- A table comparing median score of the different groups of firms.

The second part of this chapter, *Section 7.3*, presents the results of the INTERPROD *project*-level questionnaires, (INT-B, INT-C, INT-D and INT-E). As in the previous section, these data were collected from electronics firms in Ireland and the UK. Section 7.3 is divided into nine sub-sections each of which presents results on a different aspect of managing NPD projects:

1.R & D/Marketing Integration – INT-B
2.Project Management Skills – INTB,

3.Top Management Support – INT-B,
4.Development Process Proficiency – INT-B,
5.New Product Characteristics – INT-C,
6.Market Characteristics – INT-C,
7.Marketing and Technological Skills – INT-C,
8.Marketing and Technological Synergy – INT-C,
9.Methods of Organising NPD – INT-B

INTERPROD questionnaires INT-B to INT-E were filled out by respondents in relation to successful and failed NPD projects, this allowed the management of successful new products (those that met or exceeded expectations) and failed new products (those that failed to meet expectations) to be compared.

Most of the results presented in Section 7.3 are in the form of correlation coefficients measuring the relationship between different management practices and project success. The sub-sections in Section 7.3 are laid out in a similar manner to those in Section 7.2 with each sub-section containing the following:

- A brief description of the data about to be presented.
- Brief observations on the data presented.
- A table containing correlation coefficients.

In this section it is not practical to compare means as this would require comparisons between 12 different groups of projects – failed and successful projects in each of the six groups of firms defined in Table 7.1. Therefore, the comparison of data sets, described by Tests 1, 2 and 3, is conducted in the 'Observations' sub-section of Section 7.3.

The summary of this chapter, Section 7.4, provides a link forward to Chapter 8 where the results presented here are discussed in detail.

7.1.3 Reliability of Data

The reliability of a survey is an indication of how consistent it is as a measuring instrument, in other words the extent to which similar results will be recorded under repeated measurements. As discussed in Chapter 5, it is not possible to measure the reliability of the INTERPROD survey. However, previous researchers (Yap, 1992; Souder & Song, 1997; Souder et al. 1998; Souder & Jenssen, 1999) have not considered this to be a serious problem. As recommended by Yap, (1992) the reliability of INTERPROD has been increased by using clear and unambiguous questions and instructions, and also by collecting data using personal interviews which allowed misunderstandings about questions to be quickly resolved.

In this study, concerns about the reliability of the INTERPROD research instrument relate to the possibility of significant relationships between independent and dependent variables being overlooked. Yap (1992) used significance levels of 0.2 and 0.3 in order to avoid rejecting potentially significant relationships. In presenting the data in Section 7.3 of this chapter relationships that are significant at $p \leq 0.01$, 0.05, 0.1 and 0.2 will be highlighted. However, in the discussion that follows in Chapter 8 relationships that are significant at the higher levels of 0.1 and 0.2 will be viewed with greater caution than those that are significant at 0.01 and 0.05.

7.1.4 Summary of Results

A brief summary of the key findings in this chapter is given below.

Findings relating to research **Question 1**:

How do small Irish electronics firms manage the development of new products, and are the success factors that apply to small Irish firms different from those that apply to large Irish firms?

Differences were found between small and large firms in the following areas:

1. The involvement of external organisations.
2. The importance of manufacturing issues.
3. The impact of product reliability
4. The level and allocation of NPD resources.
5. The level of market growth and maturity.
6. The involvement of top management and commercial entities in NPD.
7. The use of project management techniques.

Findings relating to research **Question 1.a**:

Are the differences between small and large Irish electronics firms also seen between small and large UK electronics firms?
Many of the differences seen between small and large firms in Ireland were also seen between small and large firms in the UK with the exception of the following issues:

1. The involvement of external organisations.
2. The importance of manufacturing issues.
3. The use of project management techniques.

Key findings relating to research **Question 1.b**:
How does NPD management in the Irish electronics sector, as measured by the INTERPROD research tool, compare with that in the UK?
Differences between the Irish and UK firms related to:

1. Firm size and turnover.
2. The level of export.
3. The involvement of external organisations.

These findings will be discussed in detail in Chapter 8.

7.2 Company-Level Results

The results reported in this section come from data collected with the three INTERPROD questionnaires, INT-A, INT-F and INT-L, which are all aimed at the company (as opposed to project) level. Results are presented in the following six area:

1. NPD Context
2. Performance Measures
3. Development Process Formality
4. Involvement of External Organisations
5. Resources for NPD
6. Cycle-time Practices

These six areas were selected because they reflect the general topics on which the INTERPROD questionnaires collected data but also because they are consistent with the research model presented in Chapter 4. The consolidation of the INTERPROD data into the six areas listed above allows the data to be easily compared and contrasted and reduces almost 200 data items per firm into a set of manageable results.

Results are presented in Section 7.2.1 to 7.2.6 corresponding to the six areas above. Although the detailed discussion of these data is in Chapter 8 some general observations are made about each set of results.

7.2.1 NPD Context

Within the model presented in Chapter 4 the context, in terms of firm size and location, within which firms innovate is shown to have an impact on the manner in which NPD is managed and also on NPD success factors. In this study the data are segregated by firm location, data from Irish firms are analysed separately from data from other countries, in other words the location variable is explicitly considered. Table 7.2 shows data relating to firm size. The median scores for each of the parameters listed are given. The results of the Mann-Whitney *U* tests comparing the medians of each group of firms are presented in Table 7.3.

The concerns that firms have about the future are another indication of the context within which they operate. To gain an understanding of the issues about which respondents were most concerned a list of thirty-one potential challenges was presented and they were asked to record their degree of concern for each challenge on a scale from 1 – none to 5 – very large. Table 7.4 lists the top five concerns for each group of firms. The median scores for all the challenges in Table 7.4 are equal to 4.00. The mean scores are listed in the Table 7.4 to give and indication of the trends within each group. The data upon which Table 7.4 is based are presented in Appendix 5.

Observations on NPD Context

- UK firms were significantly larger than Irish firms in terms of number of employees, level of turnover and turnover per employee. This is an interesting finding because it implies that some of the differences between firms in the UK and Ireland could be due to firm size as well as location.
- Irish firms were significantly more dependent on export than UK firms.
- All firms shared similar concerns about potential challenges. Irish and UK firms share the same top three concerns, in the same order.

- *Shortening development cycles* and *Market development activities* were the only two key concerns shared by both sets of large firms in the study.
- *Quality and reliability of newly launched products, Controlling new product development costs, Monitoring changing customer needs and requirements* and *Reducing new product failure rates* were all cited as concerns for small firms in both Ireland and the UK.
- Large Irish firms were the only group who did not express a concern about financial resources for NPD projects.

	Irish Firms			**UK Firms**		
	All	**Small**	**Large**	**All**	**Small**	**Large**
Size [A6]	72.5	30	170	300	33	475
Turnover (Millions) [A8]	IR£3.9	IR£1.75	IR£17.6	ST£20.5	ST£2.0	ST£34.0
Turnover per Employee (Millions) [A8/A6]	IR£0.065	IR£0.06	IR£0.11	ST£0.69	ST£0.61	ST£0.76
% Export [A12]	95%	95%	97%	33%	30%	45%

Table 7.2 NPD Context, Median

Parameter	**Test 1 Small Irish vs. Large Irish firms, *p*-value**	**Test 2 Small UK vs. Large UK firms, *p*-value**	**Test 3 UK vs. Irish firms, *p*-value**
Size [A6]	<.0005[a]	<.0005[a]	<.0005[a]
Turnover (Millions) [A8]	<.0005[a]	<.0005[a]	<.0005[a]
Turnover per Employee (Thousands) [A8/A6]	.138	.095	<.0005[a]
% Export [A12]	.283	.313	<.0005[a]

[a] Statistically significant, $p \leq 0.05$, Mann Whitney U test

Table 7.3 Comparison of data set medians of NPD Context.

All Irish Firms	*All UK Firms*
1. Shortening development cycle times [A77] – **3.86** 2. Market development activities [A68] – **3.78** 3. Monitoring changing customer needs and requirements [A85] – **3.75** 3. Quality and reliability of newly launched products [A73] – **3.75** 5. Selection of human resources for product innovation [A83] – **3.67**	1. Shortening development cycle times [A77] – **4.00** 2. Market development activities [A68] – **3.79** 3. Monitoring changing customer needs and requirements [A85] – **3.77** 4. Keeping up with new technologies [A90] – **3.84** 5. Replacement of old products by new products [A82] – **3.71** 5. Awareness of emerging new technologies [A74] – **3.71** 5. Funding of development projects [A66] – **3.71** 5. Quality of new product ideas [A62] – **3.71**
Small Irish Firms 1. Quality and reliability of newly launched products [A73] – **3.91** 2. Selection of human resources for product innovation [A83] – **3.82** 3. Controlling new product development costs [A84] – **3.77** 3. Monitoring changing customer needs and requirements [A85] – **3.77** 5. Reducing new product failure rates [A80] – **3.72**	***Small UK Firms*** 1. Quality of new product ideas [A62] – **4.20** 1. Replacement of old products by new products [A82] – **4.20** 1. Monitoring changing customer needs and requirements [A85] – **4.20** 4. Controlling new product development costs [A84] – **3.80** 4. Reducing new product failure rates [A80] – **3.80** 4. Shortening development cycle times [A77] – **3.80** 4. Quality and reliability of newly launched products [A73] – **3.80**
Large Irish Firms 1. Market development activities [A68] – **4.14** 1. Shortening development cycle times [A77] – **4.14** 3. Quality of new product ideas [A62] – **3.79** 3. Market research processes for new products [A69] – **3.79** 5. Monitoring changing customer needs and requirements [A85] – **3.71**	***Large UK Firms*** 1. Shortening development cycle times [A77] – **4.05** 2. Market development activities [A68] – **3.89** 3. Funding of development projects [A66] – **3.84** 3. Awareness of emerging new technologies [A74] – **3.84** 3. Keeping up with new technologies [A90] – **3.84**

Note 1: The numbers given in bold in the table above are the mean scores for each potential challenge.
Note 2: Where several challenges had the same mean score all are listed.

Table 7.4 Top Potential Challenges [A62 – A93]

7.2.2 Performance Measures

Many different ways of measuring the success of a new product are used by both academics and practitioners (Driva et al. 2000). However most of these measures examine the success of individual products. The success of a NPD programme can be determined using a different set of measures (Cooper & Kleinschmidt, 1996). The data presented in this section reflect the overall success of product development within the firms studied.

Table 7.5 shows the turnover that firms generate from new products and also the average number of products launched and the average success rate each year (the ranges for these data are given in Appendix 5). Additionally respondents were asked to classify their product offerings and revenue based on the position of products within a four-stage product life-cycle. These data give an indication of the extent to which firms rely on newer or older products for sales revenue. The median scores of the different groups are compared in Table 7.6.

Observations on Performance Measures

- Both groups of firms reported that they generated similar levels of revenue from new products, launched similar numbers of new products and had similar success levels with new products once they were launched.
- There were no significant differences between Irish and UK firms in terms of the number of products at each of the four stages defined.
- Irish firms reported significantly lower contribution from products at the introduction stage and higher from products at the maturity stage.
- Large Irish electronics firms were more reliant on mature products than small Irish electronics firms.
- UK electronics firms got significantly higher contributions from introduction stage products and lower contributions from maturity stage products than Irish electronics firms.

Parameter	All Irish Firms	Small Irish Firms	Large Irish Firms	All UK Firms	Small UK Firms	Large UK Firms
% Turnover from new products [A21]	45%	40%	55%	60%	79%	52.5%
Average commercial success of new products launched during the past 3 years. [A22][1]	3.00 (*3.00*)[2]	3.00 (*3.05*)	3.00 (*2.93*)	3.00 (*2.67*)	3.00 (*2.40*)	3.00 (*2.74*)
Of the products launched over the last five years what percentage succeeded? [L34]	72.5%	77.5%	65%	80%	75%	80%
Average number of new products launched per year. [L36]	3.25	2.0	8.0	4.0	4.0	4.0
Percentage of products at:						
Introduction Stage [A13]	12.5%	20%	10%	15%	10%	15.5%
Growth Stage [A14]	20%	25%	20%	30%	20%	32.5%
Maturity Stage [A15]	40%	27.5%	44.3%	30%	30%	25%
Decline Stage [A16]	10%	10%	10%	20%	20%	17.5%
Contribution from products at:						
Introduction Stage [A17]	5%	5%	5%	10%	20%	10%
Growth Stage [A18]	24%	25%	21.5%	40%	40%	40%
Maturity Stage [A19]	50%	40%	62.5%	30%	30%	32.5%
Decline Stage [A20]	7.5%	5%	10%	10%	15%	10%

Note 1: 1 – Far above expectations, 2 – Slightly above expectations, 3 – Met expectations, 4 – Slightly below expectations, 5 – Far below expectations
Note 2: The figures in italics are mean scores.

Table 7.5 Performance Measures, Median

Parameter	Test 1 Small Irish vs. Large Irish firms, *p*-value	Test 2 Small UK vs. Large UK firms, *p*-value	Test 3 UK vs. Irish firms, *p*-value
% Turnover from new products [A21]	.745	.175	.393
Average commercial success of new products launched during the past 3 years. [A22]	.858	.600	.162
Of the products launched over the last five years what percentage succeeded? [L34]	.396	.366	.460
Average number of new products launched per year. [L36]	.033[a]	.803	.652
Percentage of products at:			
Introduction Stage [A13]	.635	.939	.783
Growth Stage [A14]	.781	.098	.644
Maturity Stage [A15]	.015[a]	.499	.197
Decline Stage [A16]	.987	.227	.071
Contribution from products at:			
Introduction Stage [A17]	.889	.502	.018[a]
Growth Stage [A18]	.987	.910	.169
Maturity Stage [A19]	.028[a]	.764	.025[a]
Decline Stage [A20]	1.000	.801	.114

[a] Statistically significant, $p \leq 0.05$, Mann Whitney U test

Table 7.6 Comparison of Performance Measures Medians

7.2.3 Development Process Formality

Many authors have examined the NPD process in terms of how proficiently organisations execute the different stages in the process and the impact that process proficiency has on new product success (Cooper & Kleinschmidt, 1987, Song & Montoya-Weiss, 1998). These studies have generally examined the NPD process at a project level. However, there is also a company level dimension to how the NPD process is managed, this is examined below.

The data presented in this section relate to the formality of the NPD process. The key findings are summarised in Table 7.7, (ranges are given in Appendix 5). The table shows the median response by each group to the questions asked, the median scores of the different groups are compared in Table 7.8.

Observations on Development Process Formality

- All firms in the study, small, large, Irish and UK, reported that they recorded prototype test results proficiently.
- All firms in the study reported that extensive discussions about NPD took place outside formal meetings.
- The mean scores show that large firms are more proficient at recording results, are more likely to hold post-launch meetings and are less likely to have extensive discussions outside formal meetings. Small firms reported that their product history files were less well organised than those of large firms.

	All Irish Firms	Small Irish Firms	Large Irish Firms	All UK Firms	Small UK Firms	Large UK Firms
Customer reactions to early designs are proficiently recorded.[1] [L7]	4 *(3.31)*[4]	4 *(3.05)*	4 *(3.71)*	4 *(3.44)*	4 *(3.29)*	4 *(3.50)*
Safety test results are proficiently recorded.[1] [L8]	4 *(3.81)*	4 *(3.68)*	5 *(4.00)*	5 *(3.90)*	4 *(3.14)*	5 *(3.90)*
Prototype test results are proficiently recorded.[1] [L9]	4 *(4.08)*	4 *(4.05)*	4 *(4.14)*	4 *(4.04)*	4 *(3.86)*	4 *(4.04)*
Post launch meetings are frequently conducted.[1] [L16a]	3 *(2.81)*	3 *(2.77)*	4 *(2.86)*	4 *(2.92)*	3 *(2.86)*	4 *(2.95)*
Extensive discussions about NPD take place outside formal meetings.[1] [L21]	4 *(4.06)*	4 *(4.09)*	4 *(4.00)*	4 *(4.21)*	4 *(4.29)*	4 *(4.18)*
How well organised is product history file?[2] [L29]	3 *(3.28)*	3 *(3.99)*	2 *(2.86)*	3 *(3.63)*	3 *(4.14)*	3 *(3.45)*

Note 1: 1 = Strongly Disagree, 2 = Disagree, 3 = Neutral, 4 = Agree, 5 = Strongly Agree

Note 2: 1 = Extremely well organised, 2 = Well organised, 3 = Somewhat organised, 4 = Somewhat disorganised, 5 = Extremely disorganised,.

Note 3: The figures in italics are mean scores

Table 7.7 Development Process Formality, Median

	Test 1 Small Irish vs. Large Irish firms, *p*-value	Test 2 Small UK vs. Large UK firms, *p*-value	Test 3 UK vs. Irish firms, *p*-value
Customer reactions to early designs are proficiently recorded [L7]	.601	.242	1.00
Safety test results are proficiently recorded [L8]	.111	.340	.164
Prototype test results are proficiently recorded [L9]	.912	.400	.666
Post launch meetings are frequently conducted [L16a]	.919	.443	.344
Extensive discussions about NPD take place outside formal meetings [L21]	.915	.750	.455
How well organised is product history file?[1] [L29]	.189	.812	.459

[a] Statistically significant, $p \leq 0.05$, Mann Whitney U test

Table 7.8 Comparison of Development Process Formality.

7.2.4 Involvement of External Organisations

The interaction of firms with external organisations in their NPD projects was one of the areas addressed in INTERPROD. As technology becomes more complex it is increasingly difficult for firms, particularly small firms, to have in-house all the resources necessary to develop new technology based products. Additionally, as product and technology life cycles decrease it is important for firms to engage at an early stage with both their suppliers and customers. Figures 7.2 and 7.3 show the median frequency of interaction with six different types of organisations 1) universities, 2) research institutes, 3) competitors, 4) other companies, 5) customers, and 6) suppliers. The data upon which these graphs are based is presented in a tabular format in Appendix 5. Table 7.9 compares the median scores of the small vs. large and Irish vs. UK firms.

Observations on Involvement of External Organisations

- The overall frequency of interaction with external organisations, with the exception of customers and suppliers, is quite low. Scores for the frequency of interaction were generally less than three; in terms of the scale used this indicates that most firms interacted externally less than 'occasionally'.
- Small firms in each country interacted less regularly with suppliers, than large firms. In Ireland the difference was statistically significant.
- A significant difference was also found between the frequency of interaction with customers in small and large Irish firms. This difference was not seen in the UK sample.
- There were significant differences between interaction with competitors and suppliers in UK and Irish electronics firms. In both cases Irish firms interacted less frequently than UK firms.

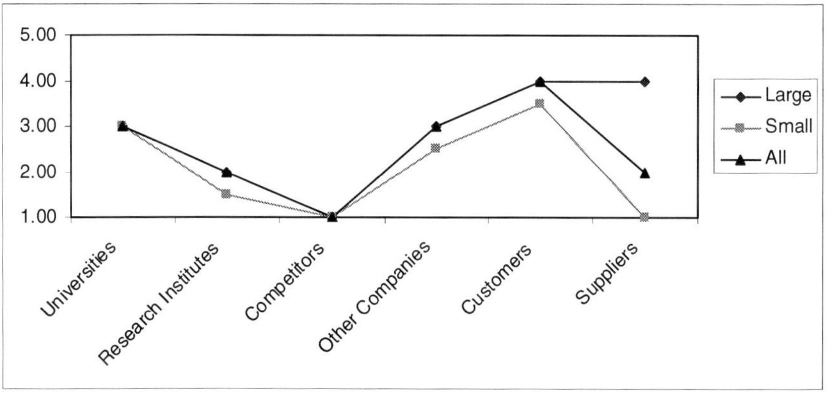

Figure 7.2 Involvement of External Organisations Irish Firms [A36 – A41]

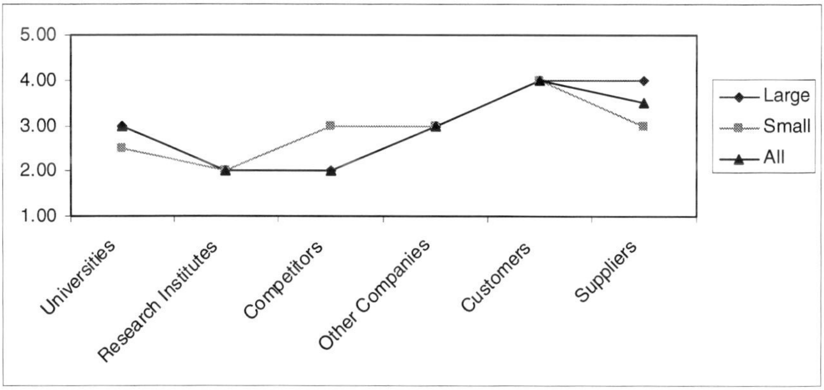

Figure 7.3 Involvement of External Organisations UK Firms [A36 – A41]

Note: A five-point Likert-type ranking scale (1 = never, 2 = seldom, 3 = occasionally, 4 = often, and 5 = very often) was used to measure the frequency of interaction with a range of external organisations.

Interaction with:	Test 1 Small Irish vs. Large Irish firms, *p*-value	Test 2 Small UK vs. Large UK firms, *p*-value	Test 3 UK vs. Irish firms, *p*-value
Universities	.757	.237	.263
Research Institutes	.167	.577	.868
Competitors	.255	.675	.002[a]
Other Companies	.212	1.00	.132
Customers	.017[a]	.867	.961
Suppliers	.003[a]	.202	.010[a]

[a] Statistically significant, $p \leq 0.05$, Mann Whitney *U* test

Table 7.9 Interaction with External Organisations – Median Values. [A36-A41]

7.2.5 Resources for NPD

Respondents were asked about the resources they allocate for NPD. Their overall R&D spending as a percentage of sales and also the percentage of employees in R&D was recorded for each group. Results are shown in Table 7.10. Respondents were presented with seven different categories and asked to divide their product innovation spending between them. Figures 7.4 and 7.5 show the median percentage expenditure in each of the seven categories. For each graph the total spending by each group sums to 100%. Table 7.11 compares the medians of the data recorded from each group of interest.

Observations on Resources for NPD

- Results indicate that NPD is a strategic activity for all the firms surveyed; all firms engage a high percentage of employees in NPD – the minimum being 14% of employees in R&D in large Irish electronics firms. Additionally, each group of firms invests a significant amount of sales revenue in NPD – again the minimum is in large Irish electronics firms.

- R&D expenditure and intensity was significantly lower in large Irish firms than in small Irish firms. Also R&D intensity was lower in Irish firms than in UK firms.
- The highest level of product innovation spending in all the firms surveyed was on new product development. The lowest spending was on basic research.
- No significant differences were found between the nature of product innovation spending in small and large firms in the UK.
- There was only one significant difference between the nature of product innovation spending in small and large Irish firms. Small Irish firms spent less on process improvement than large Irish firms.
- Several differences were found between UK and Irish electronics firms. UK firms spent more on applied research and marketing new products while Irish firms spent more on new product development.

	Irish Firms			**UK Firms**		
	All	**Small**	**Large**	**All**	**Small**	**Large**
R&D expenditure as a % of sales [A33]	11.5 %	13%	6%	11%	18%	9%
% of employees in R&D (R&D Intensity) [A30/A6]	16.5 %	26%	14%	21%	21%	15%

Table 7.10 Resources for NPD.

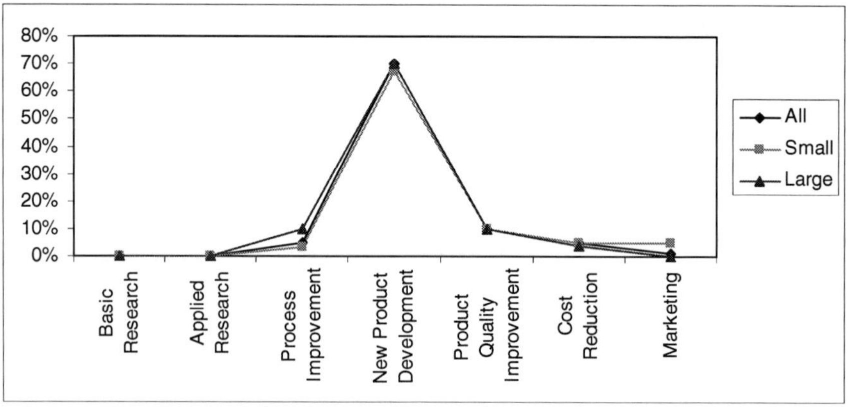

Figure 7.4 Breakdown of spending on product innovation – Irish Firms [A23 – A29].

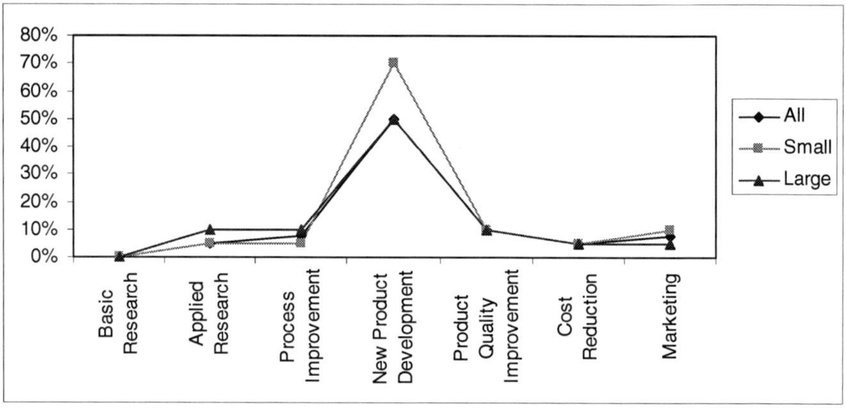

Figure 7.5 Breakdown of spending on product innovation – UK Firms [A23 – A29].

Parameter	Test 1 Small Irish vs. Large Irish firms, *p*-value	Test 2 Small UK vs. Large UK firms, *p*-value	Test 3 UK vs. Irish firms, *p*-value
R&D expenditure as a % of sales [A33]	.012[a]	.165	.976
% of employees in R&D (R&D Intensity) [A30/A6]	.031[a]	.749	<.0005[a]
Spending on:			
Basic Research [A23]	.814	.881	.992
Applied Research [A24]	.828	.204	.008[a]
Process Improvement [A25]	.033[a]	.768	.546
New Product Development [A26]	.961	.298	.004[a]
Product Quality Improvement [A27]	.853	.307	.950
Cost Reduction [A28]	.814	.455	.231
Marketing [A29]	.1076	.771	.019[a]

[a] Statistically significant, $p \leq 0.05$, Mann Whitney *U* test

Table 7.11 Comparison of data set medians of Resources for NPD.

7.2.6 Cycle Time Practices

This section presents results on product development cycle times and the techniques used to reduce these cycle times. It also looks at the involvement of users at the early stage of new product development. Some of the general findings on practices that impact on cycle times are shown in Table 7.12. The median scores of the different groups are compared in Table 7.13.

Respondents were also asked about the techniques and methods that they used to reduce new product development cycle times and the impact that these techniques and methods had on cycle times. For each technique there are two things that are of interest, firstly how many of each group used the technique, secondly how effective it was perceived to be in reducing NPD cycle time. Twenty-eight techniques were presented to the respondents. Table 7.14 shows details of the top three most commonly used techniques for each group listing the percentage of firms that used each technique. Details of the overall responses to this part of the questionnaire are given in Appendix 5.

The techniques that were found to be most effective in reducing product development cycle time are listed in Table 7.15, only techniques that were used by more than 70% of each groups are included in this analysis. The scale used to measure the effectiveness of a particular technique was:

1. Helped us significantly reduce our cycle times.
2. Helped us moderately reduce our cycle times.
3. Helped us reduce some cycle times.
4. Has not had any impact on our cycle times
5. Has actually hurt us in some ways

As with the rest of the INTERPROD data, the data presented in Table 7.15, are non-normal, however the mean scores give an indication of which techniques were found to be the most effective by the firms interviewed.

Observations on Cycle Time Practices

- Time to market was very similar in small and large Irish firms, but small UK firms developed new products significantly faster than large UK firms. Also, Irish firms took a significantly shorter time to develop new products than UK firms.
- There were only two statistically significant differences between the cycle time practices of small and large Irish firms. Small Irish firms differ from large Irish firms in that they are less likely to consider manufacturing details when they are developing new products and also they are less likely to partner with their suppliers.

- No significant differences were found between the cycle time practices of small and large UK electronics firms or between the cycle time practices of Irish and UK electronics firms.
- It is interesting to see from Table 7.13 that there is no clear trend in the differences between small and large firms in the UK and Ireland. For example, in Ireland small firms contact customers more frequently than large firms while the opposite is true in the UK.
- The only technique that appears in the top three most commonly used techniques for each group of firms is *Anticipating customer's future needs,* there is no other overlap between the top three techniques in UK and Irish firms.
- There is however a high degree of similarity between small Irish and UK firms. The top three most commonly used techniques were the same for each group; *Anticipating customer's future needs, Attention to manufacturability during design,* and *Setting clear project objectives.*
- Within the large firms, two of the top three techniques are common to UK and Irish firms; *Adoption of ISO9000 standards* and *Anticipating customer's future needs.*
- Comparing Tables 7.14 and 7.15 suggests that the techniques that are commonly used are not necessarily those that are most effective.
- The techniques that were found to be effective in reducing NPD cycle times in both sets of small firms were; *Setting clear project objectives,* and *Establishing clear priorities,* these were not effective in large firms.
- The techniques that were found to be effective in reducing NPD cycle times in both sets of large firms were; *Attention to manufacturability during design, CAD/CAM,* and *Customer inputs at the design stage.*

	All Irish Firms	Small Irish Firms	Large Irish Firms	All UK Firms	Small UK Firms	Large UK Firms
Average time from project initiation to commercialisation[1] [A34]	2.5	2.5	2.5	3	2	3
We rely on users to help us define and clarify their needs in NPD.[2] [F6]	4 *(3.83)*	4 *(3.68)*	4 *(4.07)*	4 *(3.88)*	4 *(4.00)*	4 *(3.85)*
How frequently do you contact customers about NPD?[3] [F7]	3 *(2.83)*	3 *(3.00)*	3 *(2.57)*	2.5 *(2.69)*	2 *(2.23)*	3 *(2.86)*
We think more in terms of platforms than products.[2] [F19]	4 *(3.94)*	4 *(3.86)*	4 *(4.07)*	4 *(3.75)*	4 *(3.54)*	4 *(3.82)*
Manufacturing details are considered explicitly in designing our new products.[2] [F20]	4 *(4.28)*	4 *(4.05)*	5 *(4.64)*	4 *(3.90)*	4 *(3.77)*	4 *(3.95)*
Our design changes grow more frequent as we move closer to manufacturing.[2] [F22]	3 *(2.86)*	3 *(3.05)*	2 *(2.57)*	3 *(2.87)*	3 *(2.54)*	3.5 *(2.97)*
We often partner with our suppliers.[2] [F23]	3 *(2.78)*	2.5 *(2.36)*	4 *(3.43)*	3 *(3.15)*	3 *(2.69)*	3.5 *(3.31)*

Note 1: Data presented are based on a scale; 1 – less than 6 months; 2 – 6 months to 1 year; 3 – 1 to 3 years; 4 – more than 3 years.
Note 2: Data presented show the extent of agreement with the statement given: 1 – Strongly Disagree, 2 – Disagree, 3 – Neutral, 4 – Agree, 5 – Strongly Agree
Note 3: Data presented indicate the frequency of contact based on a scale; 1 – Daily, 2 – Weekly, 3 – Monthly, 4 – Yearly, 5 – None.

Table 7.12 Cycle Time Reduction – General Findings, Median

	Test 1 Small Irish vs. Large Irish firms, *p*-value	**Test 2 Small UK vs. Large UK firms, *p*-value**	**Test 3 UK vs. Irish firms, *p*-value**
Average time from project initiation to commercialisation [A34]	.957	.006[a]	.049[a]
We rely on users to help us define and clarify their needs in NPD. [F6]	.256	.458	.411
How frequently do you contact customers about NPD? [F7]	.228	.064	.214
We think more in terms of platforms than products. [F19]	.413	.604	.551
Manufacturing details are considered explicitly in designing our new products. [F20]	.008[a]	.788	.061
Our design changes grow more frequent as we move closer to manufacturing. [F22]	.200	.228	.189
We often partner with our suppliers. [F23]	.008[a]	.231	.352

[a] Statistically significant, $p \leq 0.05$, Mann Whitney U test

Table 7.13 Comparison of Cycle Time Practices.

All Irish Firms 1. Attention to manufacturability during design [F40] – 97% 2. Anticipating customer's future needs [F39] – 94% 3. Close collaboration between R&D and manufacturing [[F58] – 94%	*All UK Firms* 1. Anticipating customer's future needs [F39] – 95% 2. Setting clear project objectives [F51] – 89% 3. Setting clear time goals [F52] – 89% 4. Improved project planning [F54] –89%
Small Irish Firms 1. Attention to manufacturability during design [F40] – 95% 2. Anticipating customer's future needs [F39] – 91% 3. Setting clear project objectives [F51] – 91% 4. Close collaboration between R&D and manufacturing [F58] – 91%	*Small UK Firms* 1. Anticipating customer's future needs [F39] – 100% 2. Attention to manufacturability during design [F40] – 100% 3. Setting clear project objectives [F51] – 100% 4. Setting clear time goals [F52] – 100% 5. Improved project planning [F54] – 100%
Large Irish Firms 1. Anticipating customer's future needs [F39] – 100% 2. Attention to manufacturability during design [F40] – 100% 3. Adoption of ISO9000 standards [F49] – 100% 4. Setting clear time goals [F52] – 100% 5. Establishing clear priorities [F53] – 100% 6. Improved project planning [F54] – 100% 7. Close collaboration between R&D and manufacturing [F58] – 100%	*Large UK Firms* 1. Anticipating customer's future needs [F39] – 93% 2. Adoption of ISO 9000 standards. [F49] – 93% 3. Setting clear project objectives [F51] – 86% 4. Setting clear time goals [F52] – 86% 5. Establishing clear priorities [F53] – 86% 6. Improved project planning [F54] – 86% 7. Clear documentation of the development process. [F55] – 86% 8. Customer inputs at the design stage [F59] – 86%

Table 7.14 Most Commonly Used Cycle Time Reduction Techniques [F39 – F66]

All Irish Firms 1. CAD/CAM [F66] – 1.88 2. Improved project planning [F54] –2.1 3. Close collaboration between R&D and manufacturing [F58] - 2.15 4. Customer inputs at design stage [F59] – 2.16 5. Simulation [F50] – 2.19 6. Establishing clear priorities [F53] –2.19	*All UK Firms* 1. Greater team responsibility (empowerment) [F41] – 2.00 2. CAD/CAM [F66] – 2.00 3. Attention to manufacturability during design [F40] – 2.20 4. Anticipating customer's future needs [F39] – 2.28 5. Setting clear project objectives [F51] – 2.35 6. Setting clear time goals [F52] – 2.35
Small Irish Firms 1. Improved project planning [F54] – 1.94 2. Developing high quality prototypes [F60] - 2.06 3. CAD/CAM [F66] – 2.11 4. Establishing clear priorities [F53] – 2.17 5. Setting clear project objectives [F51] – 2.30	*Small UK Firms* 1. Setting clear project objectives [F51] – 2.20 2. Setting clear time goals [F52] – 2.20 3. Greater team responsibility (empowerment) [F41] – 2.25 4. Anticipating customer's future needs [F39] – 2.40 5. Establishing clear priorities [F53] – 2.50 6. Close collaboration between R&D and manufacturing [F58] – 2.50
Large Irish Firms 1. CAD/CAM [F66] – 1.54 2. Close collaboration between R&D and manufacturing [F58] – 1.86 3. Customer inputs at design stage [F59] – 1.92 4. Attention to manufacturability during design [F40] – 2.00 5. Use of concurrent engineering and design [F43] – 2.00	*Large UK Firms* 1. Greater team responsibility (empowerment) [F41] – 1.90 2. Attention to manufacturability during design [F40] – 2.00 3. CAD/CAM [F66] – 2.00 4. Anticipating customer's future needs [F39] – 2.23 5. Improved project planning [F54] –2.33 6. Customer inputs at design stage [F59] – 2.33

Table 7.15 Most Effective Cycle Time Reduction Techniques, Mean

7.3 Project Level Results

Four of the INTERPROD questionnaires, INT-B to INT-E, addressed issues relating to specific new product projects within the firm. Firms were asked to complete these with respect to one successful and one unsuccessful new product. Not all firms were in a position to identify both a successful and an unsuccessful new product. Table 7.1 shows the number of successful and unsuccessful projects in the sample. By correlating management practices with the level of project success (as has been defined in Chapter 5) it is possible to propose a link between particular practices and success. This approach to analysing INTERPROD data has also been adopted by several authors who have worked with the INTERPROD data for example; Yap & Souder, 1994; Song, Souder & Dyer, 1997; Souder et al., 1997 and several other publications, as discussed in Chapter 3.

This section of the report presents findings on practices that were found to correlate with new product success (i.e. success factors), and on the differences that were found between the success factors in large and small firms. The four INTERPROD questionnaires examined in this section measure a total of 156 items. Not all of theses are presented here. Only items that are found to correlate significantly (i.e. at $p \leq 0.01$, or $p \leq 0.05$, two-tailed) with one of the six groups of firms analysed are presented.

The results are presented in the following nine areas.
1. R&D/Marketing Integration
2. Project Management Skills,
3. Top Management Support,
4. Development Process Proficiency,
5. New Product Characteristics,
6. Market Characteristics,
7. Marketing and Technological Skills,
8. Marketing and Technological Synergy,
9. Methods of Organising NPD

7.3.1 R&D/Marketing Integration

The level of integration between technical and commercial entities has been linked with new product success (Griffin & Hauser, 1996; Song & Parry, 1997; Leenders & Wierenga, 2002). In this study four variables are used to measure the level of R&D/Marketing integration, these are shown in Table 7.16.

Observations on R&D/Marketing Integration

- Two factors were found to correlate significantly ($p \leq 0.05$) with both the UK and Irish firms in the study. These were (1) *Level of contact flow between commercial and technical entities,* and (2) *Level of information flow between commercial and technical entities.*
- R&D/Marketing integration on NPD projects correlated more strongly with small than with large firms in both Ireland and the UK.

	Irish Firms			**UK Firms**		
	All	**Small**	**Large**	**All**	**Small**	**Large**
Level of contact between commercial and technical entities [B18]	$.342^a$	$.351^b$	.203	$.339^b$	$.828^a$	.117
Level of information flow between commercial and technical entities [B19]	$.389^a$	$.441^a$	.157	$.356^b$	$.594^c$	.244
Level of participation in problem solving by commercial entities [B21]	$.274^b$	$.250^d$	.121	.204	.426	.071
Level of participation in problem solving by technical entities [B22]	$.237^c$	$.291^c$	.193	$.348^b$	.353	$.398^b$

Note: Correlation coefficients statistically significant at: $^a\ p \leq 0.01$, $^b\ p \leq 0.05$, $^c\ p \leq 0.10$, $^d\ p \leq 0.20$, two-tailed

Table 7.16 R&D/Marketing Integration

7.3.2 Project Management Skills

Six variables have been used to measure the skills and authority of the project manager. Table 7.17 shows the relationship between these skills and new product success for each of the six groups of firms in this study.

Observations on Project Management Skills

- Only one factor, the participative style of the project manager, was found to correlate significantly ($p \leq 0.05$) with success in Irish firms in the study.
- The skills of the project manager were more important for UK firms than for Irish firms.

	Irish Firms			UK Firms		
	All	**Small**	**Large**	**All**	**Small**	**Large**
Technical skills of the project manager [B28]	-.021	.079	-.144	.489[a]	.304	.545[a]
Marketing skills of the project manager [B29]	.035	.074	.092	.459[a]	.138	.524[a]
Management skills of the project manager [B30]	.158	.106	.039	.556[a]	.676[b]	.514[a]
Participative style of project manager [B31]	.287[b]	.214	.325[d]	.403[b]	.570[c]	.361[c]
The ability of the project manager to motivate project members. [B32]	.056	.008	.058	.329[b]	.539[d]	.310[d]
Project manager had the necessary authority from top management [B33]	.237[c]	.248[d]	.052	.226[d]	.501[d]	.228[d]
The degree of detailed planning and control on this project was very high [E11]	.191[d]	.029	.092	.334[b]	.586[c]	.292[d]
The motivation of project personnel on this project was very high [E14]	.282[b]	.330[b]	.066	.459[a]	.557[c]	.426[b]

Note: Correlation coefficients statistically significant at: [a] $p \leq 0.01$, [b] $p \leq 0.05$, [c] $p \leq 0.10$, [d] $p \leq 0.20$, two-tailed

Table 7.17 Project Management Skills

7.3.3 Top Management Support

Respondents to the questionnaires were asked to indicate the extent to which top management was supportive of the NPD project; this was recorded on a scale of - from 1 - Very unsupportive to 5 – very supportive. Data was also collected about the timing of top management involvement (1 – Beginning of project to 5 – End of project).

Observations on Top Management Support

- Top management support is more important for small firms than for large firms.
- Early top management involvement is linked with success in Irish firms while later involvement is linked with success in UK firms.

	Irish Firms			**UK Firms**		
	All	**Small**	**Large**	**All**	**Small**	**Large**
Top Management support [B23]	$.392^{a}$	$.504^{a}$	.180	-.044	$.479^{d}$	$-.311^{c}$
Timing of Top Management Support [B24]	-.159	$-.235^{d}$	-.085	.043	.039	.034

Note: Correlation coefficients statistically significant at: $^{a}\ p \leq 0.01$, $^{b}\ p \leq 0.05$, $^{c}\ p \leq 0.10$, $^{d}\ p \leq 0.20$, two-tailed

Table 7.18 Top Management Support

7.3.4 Development Process Proficiency

The INTERPROD project concentrates on the new product development process. New product development is one of many processes that make up product innovation, it has been found to be the product innovation; process that attracts most spending in the high-technology sector examined by the INTERPROD project (see Figures 7.4 and 7.5). For the purpose of this study the product development process was divided into the eight stages listed below:

1. *Exploratory:* Search and inquiry activities, usually phenomenon oriented.
2. *Concept Development*: Concept elaboration, extension and substantiation activities aimed at the clarification or elaboration of previously generated ideas or concepts
3. *Prototype Development:* Differs from stage 2 in that a commercially relevant prototype, first model or product has been identified and is targeted.
4. *Prototype Testing*: Laboratory, field or production-scale evaluations
5. *Market Development*: Market generation, demand simulation and market analyses activities.
6. *Manufacturing Start-Up*: Initial production runs, scale-up and preparation for full-scale activities.
7. *Marketing Start-Up*: Preparation for full-scale market entry.
8. *Technical Service*: Follow-on market and technical activities which accompany the introduction of the new product.

In the majority of NPD projects these eight stages tend to take place in the order presented above. However, the process of developing a new product is not linear; in most cases there will be one or several loops back to a previous stage or stages; it is also common for stages to occur in parallel. Respondents were asked to estimate the percentage of NPD time spent at each of these eight stages. This was not always easy to estimate given the nature of the NPD process but in all cases respondents were able to decide where the time had been spent in developing the new product in question.

Results are summarised in Figures 7.6, and 7.7. The data from which these graphs have been developed is summarised in Appendix 5. Note, the total time allocated across all the stages adds up to 100% for each firm; however the sum of the medians does not necessarily equal 100%. Table 7.19 compares the median time spent by different groups of firms at each stage of the process.

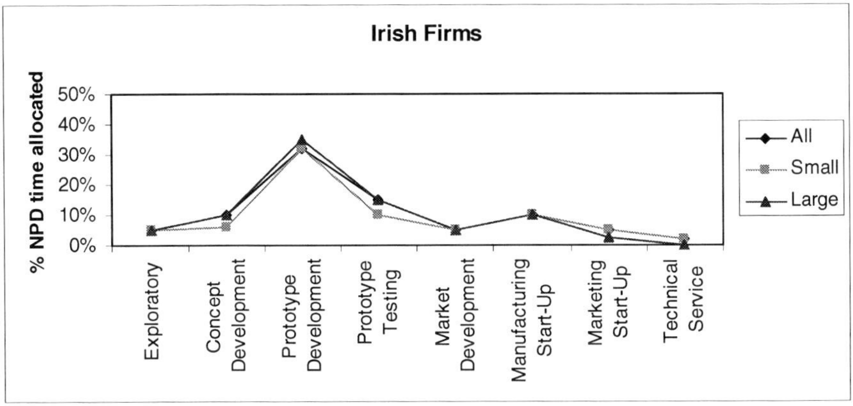

Figure 7.6 Allocation of time to NPD stages - Irish Firms [B34 – B41]

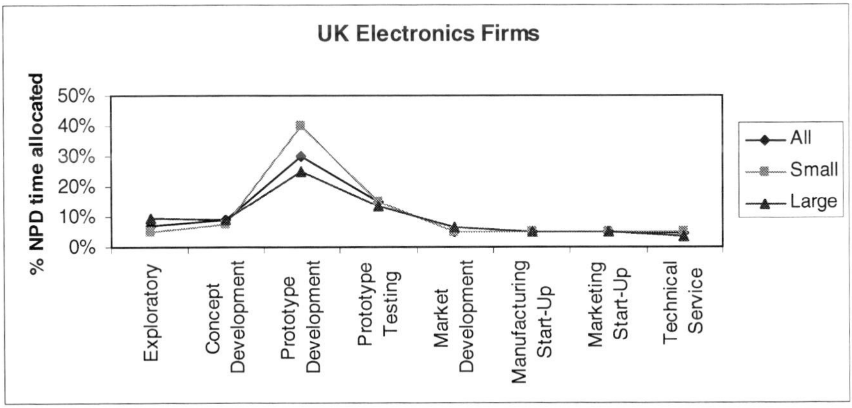

Figure 7.7 Allocation of time to NPD stages – UK Firms [B34 – B41]

Stage:	Test 1 Small Irish vs. Large Irish firms, *p*-value	Test 2 Small UK vs. Large UK firms, *p*-value	Test 3 UK vs. Irish firms, *p*-value
Exploratory [B34]	.669	.410	.093
Concept Development [B35]	.646	.429	.149
Prototype Development [B36]	.883	.030[a]	.145
Prototype Testing [B37]	.515	.725	.602
Market Development [B38]	.994	.399	.462
Manufacturing Start-Up [B39]	.271	.689	.005[a]
Marketing Start-Up [B40]	.159	.378	.792
Technical Service [B41]	.324	.219	.428

[a] Statistically significant, $p \leq 0.05$, Mann Whitney *U* test

Table 7.19 Comparison of medians of time allocation to NPD stages. [B34 – B41]

It is also interesting to look at the above data in terms of the time spent at *Pre-Development Activities* (Exploratory and Concept Development), *Development Activities* (Prototype Development and Prototype Testing) and *Product Introduction* (Marketing Development, Manufacturing Start-Up, Marketing Start-Up and Technical Service). The differences between the median distribution of time between these three types of product development activities in both successful and unsuccessful projects are summarised in Table 7.20.

All Firms	**Successful (n = 61)**	**Unsuccessful (n = 41)**	***p*- value**
Pre-Development	15%	11%	.024[a]
Development	50%	60%	.841
Product Introduction	25%	25%	.034[a]
Small Firms	**Successful (n = 26)**	**Unsuccessful (n = 21)**	***p*- value**
Pre-Development	14.5%	10%	.358
Development	46.5%	64%	.013[a]
Product Introduction	35%	24%	.110
Large Firms	**Successful (n = 35)**	**Unsuccessful (n = 20)**	***p*- value**
Pre-Development	20%	15%	.036[a]
Development	50%	55%	.695
Product Introduction	25%	30%	.075

[a] Statistically significant, $p \leq 0.05$, Mann Whitney *U* test

Table 7.20 Time allocation in successful and unsuccessful projects

In addition to recording the time spent at each stage in the NPD process, respondents were also asked to evaluate their performance at each of the 8 stages by rating it on a scale from 1 = Very Poor to 5 = Very Good. The recorded proficiency in each of the 8 stages was then correlated with the level of project success in order to assess the impact of the product development process on the success of new products. Table 7.21 shows the correlation between the proficiency at different activities and new product success.

Observations on Development Process Proficiency

- The general trend is for firms to spend most of the total product development time at prototype development and testing.

- No significant differences were found between the median % of time spent on the different stages of the NPD process in small and large Irish firms.
- In the UK, small firms were found to spend longer at prototype development than large firms.
- One significant difference in time allocation was found between UK and Irish electronics firms. The Irish firms spent longer at manufacturing start-up than the UK firms.
- A differentiator between successful and failed products would appear to be the time spent at the predevelopment and product introduction activities. However, in small firms the amount of time spent at product development activities differentiated between success and failure, while in large firms the time spent at predevelopment activities was found to differentiate between success and failure.
- Proficiency in manufacturing start-up is significant for large Irish and small UK firms but correlates negatively with success in small Irish firms.
- Prototype development and market development were the most critical development process factors for small Irish firms.

Stage	All Irish Firms	Small Irish Firms	Large Irish Firms	All UK Firms	Small UK Firms	Large UK Firms
Proficiency in the Exploration [B50]	.188^{d}	.238^{d}	.132	.342^{b}	.000	.423^{b}
Proficiency in Concept Development [B51]	.249^{b}	.402^{b}	.106	.257^{d}	-.036	.336^{c}
Proficiency in Prototype Development [B52]	.425^{a}	.423^{a}	.302^{d}	.441^{a}	.460^{d}	.494^{a}
Proficiency in Prototype Testing [B53]	.290^{b}	.260^{d}	.148	.327^{b}	.233	.394^{b}
Proficiency in Market Development [B54]	.301^{b}	.423^{a}	.272^{d}	.418^{b}	.443^{d}	.464^{b}
Proficiency in Manufacturing Start-Up [B55]	.155	-.010	.419^{b}	.319^{c}	.721^{b}	.221
Proficiency in Marketing Start-Up [B56]	.378^{a}	.388^{b}	.388^{c}	.522^{a}	.623^{c}	.500^{a}
Proficiency in Technical Service [B57]	.264^{b}	.264^{d}	.173	.059	.409	-.065

Note: Correlation coefficients statistically significant at: ^{a} $p \leq 0.01$, ^{b} $p \leq 0.05$, ^{c} $p \leq 0.10$, ^{d} $p \leq 0.20$, two-tailed

Table 7.21 Development Process Proficiency [B50 – B57]

7.3.5 New Product Characteristics

Many studies of the important factors in successful new product development have suggested that product superiority is the key to successful products (Cooper, 1990; Karakaya & Kobu, 1994; Yap & Souder, 1994). This study also investigated the characteristics of the products developed. Table 7.22 summarises the new product characteristics that had an impact on the successful outcome of the projects.

Observations on New Product Characteristics

- Product reliability and compatibility were positive selling points for small firms but was negatively correlated with success in all of the large electronics firms.
- A competitive, initial cost was the most important new product characteristic for large Irish firms but was not significant for any other group of firms in the study.
- Cost effectiveness was a more significant characteristic for large firms than for small firms.

	Irish Firms			UK Firms		
	All	Small	Large	All	Small	Large
A competitive initial cost was a planned selling point for this product. [C37]	.297^{b}	.170	.426^{b}	-.079	-.170	-.147
Cost effectiveness was a planned selling point for this product. [C42]	.177^{d}	.126	.249	.311^{c}	.017	.404^{b}
Service and Support was a planned selling point for this product [C43]	.209^{d}	.003	.401^{b}	.192	.541^{d}	.047
Ease of Implementation was a planned selling point for this product [C44]	.073	.055	.090	.302^{c}	.340	.294^{d}
Product Reliability was a planned selling point for this product [C45]	.021	.159	-.155	-.076	.340	-.167
Product Compatibility as a planned selling point for this product [C46]	.120	.339^{b}	-.220	-.104	.097	-.191
This product was positioned based on specific product features (e.g. light weight) [C24]	.190	.160	.243	.358^{b}	-.070	.492^{a}
This product was positioned based on the benefits and solutions it provided to customers [C25]	.246^{c}	.284^{c}	.264^{d}	.474^{a}	.491	.509^{a}

Note: Correlation coefficients statistically significant at: $^{a}\ p \leq 0.01$, $^{b}\ p \leq 0.05$, $^{c}\ p \leq 0.10$, $^{d}\ p \leq 0.20$, two-tailed

Table 7.22 New Product Characteristics

7.3.6 Market Characteristics

The market into which a new product is launched has been found to impact on the product's level of success (Yap & Souder, 1994). Data was collected about the potential customer base and the market for new products Table 7.23 shows the relationships between this data and new product success.

Observations on Market Characteristics

- Quickly growing markets were found to favour large firms but not small firms.
- Mature markets were negatively correlated with success for large firms but slightly positively correlated with success for small firms.
- Irish firms, and particularly small Irish firms, were more successful with new products for which their customers had a great need while UK firms, and (again) particularly small UK firms, were more successful with new products when they had a good relationship with the customers.

	Irish Firms			UK Firms		
	All	Small	Large	All	Small	Large
The customer had a great need for this type of product [C8]	.325^a	.419^a	.192	.188	.269	.170
The customers had a good relationship with us [C11]	.132	.097	.162	.427^a	.703^b	.311^c
The customers were satisfied with existing, competitors' products [C12]	-.141	-.170	-.123	-.345^b	-.917^a	-.207
The potential customers were eager to take on new products [C13]	.297^b	.285^c	.258	.216^d	-.039	.316^d
We took into consideration customers' problems with previous products when we designed this product [C23]	.456^a	.559^a	.337^c	.263^d	.284	.227
The market for this product was growing quickly [C14]	.336^a	.210	.529^a	.282^c	.000	.410^b
The market for this product was in the mature stage [C15]	-.086	.032	-.206	-.173	.283	-.437^b
The timing and unveiling of our product concept was too late [C20]	-.128	-.288^c	.083	-.468^a	-.609^c	-.420^b

Note: Correlation coefficients statistically significant at: ^{a} $p \leq 0.01$, ^{b} $p \leq 0.05$, ^{c} $p \leq 0.10$, ^{d} $p \leq 0.20$, two-tailed

Table 7.23 Market Characteristics

7.3.7 Marketing and Technological Skills

The link between marketing and technological skills and new product success has been established (Zirger & Maidique, 1990). Although the relative importance of different skills depends on the context within which new products are developed (Wren et al. 2000). Table 7.24 presented the correlation coefficients linking marketing and technological skills with new product success.

Observations on Marketing and Technological Skills

- Marketing skills are strongly correlated with success in small but not large Irish firms.
- Technical skills, capabilities and resources are more strongly correlated with success in the UK firms than in the Irish firms.
- In both the UK and Ireland, the *availability* of resources was more important to small firms while the *quality* of resources was more important to large firms.
- In general, marketing skills appear to be linked more strongly with success than are technical skills.

	Irish Firms			UK Firms		
	All	**Small**	**Large**	**All**	**Small**	**Large**
The user's requirements were well understood [E6]	$.373^a$	$.474^a$	.076	$.352^b$	.073	$.426^b$
Market Research skills were ideal for the product [C48]	$.419^a$	$.437^a$	$.372^c$	$.322^b$	.187	$.306^d$
Sales and Marketing skills were ideal for this project [C49]	$.401^a$	$.430^a$	$.350^c$	$.367^b$	-.033	$.427^b$
Engineering skills were at the desired level for this project [D11]	.148	$.243^d$	.049	$.376^b$	.212	$.478^a$
Manufacturing skills were at the desired level for this project [D12]	$.182^d$	-.071	$.397^b$	.145	.268	.175
The level of available resources was high [E12]	.114	.149	-.092	$.281^c$	$.518^d$	.239
The quality of available resources was very adequate [E13]	$.232^c$	.131	.204	$.545^a$	.318	$.660^a$

Note: Correlation coefficients statistically significant at: $^a\ p \leq 0.01$, $^b\ p \leq 0.05$, $^c\ p \leq 0.10$, $^d\ p \leq 0.20$, two-tailed

Table 7.24 Marketing and Technological Skills

7.3.8 Marketing and Technological Synergy

The alignment of existing skills with the requirements for new product projects has been found to have an impact on project outcomes (Song & Parry, 1999). Table 7.25 shows the correlations between marketing and technological synergy and new product success.

Observations on Marketing and Technological Synergy

- Neither marketing nor technological synergy was linked with success in large Irish firms or small UK firms.
- Marketing synergy was most closely linked with success in small Irish firms.
- Both marketing and technological skills were linked with success in large UK firms.

	Irish Firms			UK Firms		
	All	Small	Large	All	Small	Large
There was close fit between existing markets and the market for this product. [C52]	$.259^b$	$.388^b$	.072	$.439^a$	.274	$.489^a$
There was close fit between the existing product line and this product. [C53]	$.333^a$	$.384^b$	.215	$.318^b$	.032	$.446^b$
There was a close fit between the firms marketing skills and the needs of this project [C54]	$.310^b$	$.468^a$	.125	$.389^b$	.328	$.378^b$
There was a close fit between the firms R&D skills and the needs of this project [D13]	.069	.107	.034	$.420^a$	.369	$.481^a$
There was a close fit between the firms engineering skills and the needs of this project [D14]	$.187^d$	$.245^d$	.170	$.326^b$	.212	$.406^b$

Note: Correlation coefficients statistically significant at: $^a\ p \leq 0.01$, $^b\ p \leq 0.05$, $^c\ p \leq 0.10$, $^d\ p \leq 0.20$, two-tailed

Table 7.25 Marketing and Technological Synergy

7.3.9 Methods of Organising NPD

Lee et al. (2000) examined the link between different methods of organising NPD and new product success – they found few significant results. Their analysis was repeated on the Irish data – again the findings were not significant. Table 7.26 shows the number of Irish and UK firms that used different organisational approaches. The most popular approaches were examined and no significant links with new product success were found. Two other organisational parameters were also examined; results are presented in Table 7.26.

Observations on Methods of Organising NPD

- The most popular method of organising NPD in Irish and UK firms is under a NPD manager. Top management control is more common in small and Irish firms, while using a matrix organisation is more popular in large firms and in UK firms. There was no significant correlation between any particular method and new product success.
- A formal NPD procedure was linked with success in small UK firms but in large Irish firms less formal procedures were linked with success.
- Increased participation in decisions is positively related to success in UK and Irish firms.

	Irish Firms			UK Firms		
	All	Small	Large	All	Small	Large
How was the project organised [B17]						
NPD Department	24	10	14	13	2	11
NPD Manager	25	13	12	19	4	15
Venture Team	13	11	2	5	0	5
Task Force	11	7	4	7	4	3
NPD Committee	5	0	5	6	2	4
Matrix Organisation	9	2	7	11	1	10
Top Management Control	20	15	5	9	3	6
Dyad	6	4	2	9	2	7
The procedure for carrying out the tasks on this project was formal [B13]	.070	.026	-.141	$.268^{c}$	$.482^{d}$	.121
Adequate participation in the decision making by the people working on the project [B15]	$.197^{d}$	.092	.134	$.277^{c}$	.124	$.351^{c}$

Note: Correlation coefficients statistically significant at: $^{a}\ p \leq 0.01$, $^{b}\ p \leq 0.05$, $^{c}\ p \leq 0.10$, $^{d}\ p \leq 0.20$, two-tailed

Table 7.26 Methods of Organising NPD

7.4 Conclusion

The results from the INTERPROD research project have been presented in this chapter. Results were presented in two parts. The first part, presented in Section 7.2, examined the company-level issues; some general observations were made on the data, in particular exploring the differences between small and large, Irish and UK firms. The second set of results was obtained by examining the project-level INTERPROD data; this was presented in Section 7.3, again differences between small and large, Irish and UK firms emerged.

The purpose of this study is to provide answers to the three research questions posed at the start of this chapter. The key findings from the analysis of the Irish and UK INTERPROD data will be grouped based on the research question to which they provide answers. A summary of these findings has already been provided in Section 7.1.4. The key differences found between the groups of firms compared in this chapter are shown in Table 7.27.

At a more general level the data presented in this chapter have shown that small firms, within the electronics sector, are heavily involved in NPD and with a high degree of success. Additionally, it has emerged quite clearly that the success strategies that apply to large firms do not necessarily apply to small firms. The differences between the UK and the Irish data sets indicate that national setting (one aspect of the external environment of the firm) has an impact on the appropriate success strategies for NPD.

These results as presented have shown some interesting differences between NPD in small and large firms in the Irish electronics sector. However to determine the impact and gain a better insight into the meaning of these results it is important to examine them in detail with reference to previously published material on the management of NPD. Also, a lot of data have been presented much of it interrelated. To get a clear, more comprehensive understanding of the issues surrounding the management of NPD in small electronics firms further analysis is necessary.

Issues relating to Test 1 Small vs. Large Irish Firms	Issues relating to Test 2 Small vs. Large UK Firms	Issues relating to Test 3 Irish vs. UK Firms
1. The involvement of external organisations. 2. The importance of manufacturing issues. 3. The impact of product reliability 4. The level and allocation of NPD resources. 5. The level of market growth and maturity. 6. The involvement of top management and commercial entities in NPD. 7. The use of project management techniques.	1. The impact of product reliability 2. The level and availability of NPD resources. 3. The level of market growth and maturity. 4. The involvement of top management and commercial entities in NPD.	1. Firm size and turnover. 2. The level of export. 3. The involvement of external organisations.

Table 7.27 Summary of Key Differences between Irish-UK and Small-Large Firms.

The main aims of Chapter 8 are (1) to examine the results presented in Chapter 7 in more detail; specifically to place them in the context of previously published results relating to the management of NPD, and (2) to analyse the INTERPROD data more thoroughly in order to gain a clearer picture of the key differences differentiating the management of NPD in small and large Irish electronics firms.

CHAPTER EIGHT

INTERPROD DISCUSSION IRISH AND UK DATA

8.1 Introduction

Chapter 8 presents a detailed discussion of the INTERPROD results. The structure of this chapter mirrors that used to present the data in Chapter 7 so that;

- Section 8.2 discusses the *company* level data presented in Section 7.2, and
- Section 8.3 discusses the *project* level data presented in Section 7.3.

Section 8.4 summarises the INTERPROD findings and identifies the issues to be explored in more detail in the three case studies that are the subject of Chapters 9 and 10. The conclusion of this chapter is in Section 8.5.

8.1.1 Process Employed

One of the key challenges in analysing the INTERPROD data is to distil the wide variety of results down to a number of key conclusions. This is made more difficult by the fact that the INTERPROD data have been collected from small and large, Irish and UK firms, and in the case of project level data, successful and unsuccessful projects. The process of extracting meaning from these data was begun in Chapter 7 where decisions were made about how to organise the results. Six areas for discussion were defined for the company level data and nine areas for the project level data as shown in Table 8.1 below.

Company Level Results – Section 8.2	**Project Level Results – Section 8.3**
1. NPD Context 2. Performance Measures 3. Development Process Formality 4. Involvement of External Organisations 5. Resources for NPD 6. Cycle-time Practices	1. R&D/Marketing Integration 2. Project Management Skills, 3. Top Management Support, 4. Development Process Proficiency, 5. New Product Characteristics, 6. Market Characteristics, 7. Marketing and Technological Skills, 8. Marketing and Technological Synergy, 9. Methods of Organising NPD

Table 8.1 Organisation of INTERPROD data.

Within this chapter the observations listed in Sections 7.2 and 7.3 of Chapter 7 are discussed in terms of the general body of literature presented in Chapter 2. At the end of each sub-section results that are useful in answering the research questions are summarised as 'Key Findings'. Results are also summarised at the end of each section so that Section 8.2.7 – Summary of Company Level Results and Section 8.3.10 – Summary of Project Level Results, start to build more detailed views of how NPD is managed in small and large, Irish and UK electronics firms.

Section 8.4 contains a summary of all the results presented in this chapter. Findings are gathered according to the research question, repeated below, which they address.

How do small Irish electronics firms manage the development of new products, and are the success factors that apply to small Irish firms different from those that apply to large Irish firms?

a) *Are the differences between small and large Irish electronics firms also seen between small and large UK electronics firms?*
b) *How does NPD management in the Irish electronics sector, as measured by the INTERPROD research tool, compare with that in the UK?*

Findings are then sorted indicating which are to be examined in more detail in the case studies and which are beyond the scope of this study. The aim of this thesis is to provide answers to the primary research question therefore the issues that will be explored further in the case studies presented in Chapter 9 are those that are directly related to the primary research question stated above. The secondary research questions have been included to explore the degree to which the results from the Irish data can be generalised beyond the Irish setting and to place the Irish findings within an international context. Finally, Section 8.4 will present an updated version of the research model based on the findings discussed in this chapter.

8.2 Company Level Results

The analysis and discussion of company level results is divided into six sections:

1. NPD Context – Section 8.2.1
2. Performance Measures – Section 8.2.2
3. Development Process Formality – Section 8.2.3
4. Involvement of External Organisations – Section 8.2.4
5. Resources for NPD – Section 8.2.5
6. Cycle-time Practices – Section 8.2.6

Key findings are summarised at the end of each section and Section 8.2.7 provides a summary of all the company level results.

8.2.1 NPD Context

Significant differences in size and turnover were found between small and large firms in Ireland and the UK. This is reassuring as it indicates that size is a valid differentiator between the two groups of firms. However, significant differences were also found between the size and turnover of the UK and Irish firms with the Irish firms being smaller by a factor of 4 and reporting a median turnover that was 5 times smaller than that reported by the UK firms. Also, when discussing these data it is useful to note that there are no statistically significant size differences between small Irish and small UK firms but that large UK firms are significantly larger than large Irish firms.

This finding is important – if, as suggested, firm size is a key issue in the management of NPD projects, then differences between UK and Irish electronics firms are as likely to be due to size as location. Table 7.1, in Chapter 7, shows that large firms dominate the UK sample, 80% of the UK electronics firms have more than 100 employees. The Irish sample is more evenly divided between small and large firms with only 40% of the Irish electronics firms employing more than 100 employees. This difference in the structure of both samples is a limitation of the research and is largely due to the focus of each study. The Irish INTERPROD data were collected specifically to explore the differences between small and large firms while the UK data were collected to examine the high technology sector generally. The fact that the UK electronics firms are significantly larger than the Irish electronics firms means that whenever significant differences are found between the UK and Irish samples the data must be examined in more detail to determine whether these differences are due to size or location. This will be done by looking at the differences between small Irish firms vs. small UK firms and at the differences between large Irish firms vs. large UK firms.

One point of interest about the small firms in this study relates to the age of the firms involved. Some of the literature on small firms uses firm size as a proxy for firm age (Walsh et al. 2002) and frequently authors link small firms with New Technology Based Firms (NTBFs), (Bessant, 1998). But in this study, as shown in Table 8.2 there is no significant difference between the ages of the small and large firms.

	Small Firms	**Large Firms**	**All Firms**	**Level of significance**
Year of establishment (Median, Range)	1984 (1973, 1991)	1986 (1973, 1995)	1985 (1973, 1995)	0.182
Median age when interviewed	11.5 years	13.5 years	12.5 years	

Table 8.2 Firm Age

Irish electronics firms were found to be significantly more dependent on export than UK electronics firms. This is a reflection of Ireland's small open economy and is representative of the broad industrial sector (NACE

30 – 33) involved in this study that was responsible for 40% of Ireland's exports in 2000. Firm size had no impact on the level of export.

The other measure of the context within which firms operate presented was the future challenges about which firms were most concerned. There was a high level of consensus between all the firms involved in the study about which challenges were of the most concern. The top challenges for each group of firms are summarised in Table 8.3.

Challenge	Firms listing this challenge amongst its top 5					
	All Irish	**All UK**	**Small Irish**	**Small UK**	**Large Irish**	**Large UK**
Shortening development cycle times	X	X		X	X	X
Monitoring changing customer needs and requirements	X	X	X	X	X	
Market development activities	X	X			X	X
Quality of new product ideas		X		X	X	
Quality and reliability of newly launched products	X		X	X		
Controlling new product development costs			X	X		
Reducing new product failure rates			X	X		

Table 8.3 Top Potential Challenges

The top two concerns of the firms surveyed, *Shortening development cycle times* and *Monitoring changing customer needs and requirements* are not unexpected. Reducing NPD cycle time is frequently cited as a major challenge for manufacturing firms (Ali et al. 1995; Dröge et al. 2000). Staying close to customers has also been identified as a key concern (Cooper, 1994).

It is interesting to note that *Market development activities* was included as one of the top three concerns of all the large firms in the study (Irish and UK electronics). However, this was not included in the top concerns of any group of small firms. On the other hand *Quality and reliability of newly launched products, Controlling NPD costs*, and *Reducing new product failure rates* were importance potential challenges for all small firms in the study but not for the large firms. Roper (1996) also found, in a study that included 385 small Irish manufacturing firms, that raising product quality was the most important objective for product innovation.

These results would suggest that while large firms are concerned about future opportunities; new product ideas, new markets, improved time-to-market; small firms are still concerned about operational issues; controlling costs, improving quality and reducing failure. These differences will be explored further in the case studies in Chapter 9.

Based on the above analysis an overview of the context within which each group of firms in the study operates is shown in Table 8.4.

Small Irish Firms Small Irish firms are significantly smaller than large Irish firms, they are highly dependent on export, and they are still concerned about NPD operational issues such as controlling costs, improving quality and reducing failure.	***Small UK Firms*** Small UK firms are significantly smaller than large UK firms, they are also concerned about NPD operational issues such as controlling costs, improving quality and reducing failure.
Large Irish Firms Large Irish firms are significantly smaller than large UK firms, they are highly dependent on export, and they are concerned about future NPD opportunities: new product ideas, new markets, improved time-to-market.	***Large UK Firms*** Large UK firms are larger than any other groups of firms in the study and they are concerned about future NPD opportunities: new product ideas, new markets, improved time-to-market.

Table 8.4 NPD Context

The key findings on *NPD Context* are summarised below:

1. The UK firms involved in this study were significantly larger than the Irish firms in terms of both size and turnover.
2. Large firms are concerned about future opportunities - new product ideas, new markets, and improved time-to-market. While small firms are still concerned about operational issues - controlling costs, improving quality and reducing failure.

8.2.2 Performance Measures

One of the most frequently used company level measures of new product performance is the turnover generated from new products (Cooper, 1996). In Ireland small firms generated less turnover from new products, 40%, than large firms, 55%. These figures are slightly lower than those published by the EU but they show a similar trend – small firms generate less of their income from new products than large firms[1]. Large UK firms generate 52.5% of turnover from new products, but 79% of turnover came from new products in small UK firms. The small number of small firms in the UK sample, 5, could account for this unusually high result.

Additional performance related data were collected from the Irish firms involved in the study. This examined to the number of patents held by firms and also the number of product variants and totally new products launched by firms per annum. These results are summarised in Table 8.5 below.

Large Irish firms launch significantly more new products per annum than small Irish firms, they launch more totally new products but the also launch many more product variants. The data in Table 8.5 show that the ratio of totally new products to product variants is 1:1.5 in small firms but is close to 1:3 in large firms. This suggests that large firms launch more new products than small firms and they are also more adept at leveraging several new product variants from one new product, an indication of a more strategic approach to NPD. Alternatively it could be an indication that large Irish electronics firms undertake a large amount of incremental product innovation.

---

[1] In 1996 the average percentage of turnover generated from new products by EU electronics firms was reported to be 45% while that in Ireland was higher at 69%.

Parameter	All Irish Firms	Small Irish Firms	Large Irish Firms	Comparison of Medians (Mann Whitney *U* test)
Number of new products (not variants of existing products) launched during the last year. [IR20]	2.25 (0, 63)	2.0 (0, 63)	4.0 (0, 24)	.016[a]
Number of new product variants launched during the last year. [IR21]	4.0 (0, 56)	3.0 (0, 32)	11.5 (0, 56)	.021[a]
Number of patents held. [IR9]	3.0 (0, 100)	1.5 (0, 17)	7.5 (0, 100)	.001[a]

[a] Statistically significant, $p \leq 0.05$.

Table 8.5 Additional Irish Data (Medians).

The number of patents held by the large Irish firms was 5 times higher than that for the small Irish firms. The reluctance of small firms to protect new products by patenting has been reported previously and is an issue that is being addressed by the Irish development agencies (Cox, 1999). Also, the small number of patents held by the Irish firms is not surprising given the low level of patent applications in Ireland in general. However, it would be incorrect to attach too much importance to the patent data. The use of patents as a measure of product innovation output, particularly in the electronics industry, is questioned by several authors (Acs & Audretsch, 1988; von Hipple, 1988).

The composition of a firm's product portfolio in terms of the newness of the products on offer is another indicator of the firm's product development performance. In large Irish firms introduction and growth stage products accounted for only 30% of product volume. In small Irish firms introduction and growth stage products accounted for 45% of product volume. But Table 7.3 in Chapter 7 shows that large firms launch 8 new products per year compared with 2 per year in small firms. These two pieces of information show that large firms have much bigger product portfolios than small firms. Eight new products in large firms with 195 active products (4%) has a smaller impact than 2 new products for small firms whose number of active products is 10.5 (19%).

No significant differences were found between UK and Irish firms in terms of the percentage of products at each stage in the cycle. But differences were found in terms of the revenue generated by products at different stages. Irish electronics firms generated less revenue from products at the introduction stage, (UK - 10%, Ireland – 5%) and more from products at the maturity stage, (UK - 30%, Ireland – 50%) than UK electronics firms. Investigating this in more detail shows that there is no significant difference between small Irish and UK firms in terms of revenue from products at different stages of the product life cycle[2] but that there are differences between large UK and Irish firms[3].

When these last two points are considered two interesting points emerge:

(a.) Large Irish electronics firms launch more new products each year, 8, than any other group of firms in the study yet they generate the least amount of revenue from products at the introduction stage. This could be because their new products take a long time to build up a significant level of sales volume – however this is unlikely in a fast moving industry like electronics. A more likely explanation is that the products that they are launching are incremental improvements that have only a minor impact on their turnover.

(b.) Small Irish electronics firms launch fewer new products each year, 2, than any other group of firms in the study and yet there is no significant difference between the revenue that they generate from products at the introduction stage and that generated by small UK electronics firms. This suggests that every new product launched by small Irish electronics firms is critical to their overall performance.

Based on the data in Table 7.3 it is difficult to say which group of firms is the most successful at NPD. Large UK firms report the highest NPD success rate, small UK firms report the highest revenue from new products, small Irish firms report the highest level of satisfaction with

---

[2] The statistical significance of the differences between introduction $p = .214$; maturity $p = .470$.

[3] The statistical significance of the differences between introduction $p = .010$; maturity $p = .002$.

their NPD projects, and large Irish firms launch the highest number of new products per year. This is a very clear indication of how difficult it is to compare NPD success between firms. In keeping with previous researchers (Pillai et al., 2002; Griffin & Page, 1996) the measure of success that will be used in the rest of this section is the degree to which the commercial success of new products met the expectations of the firm. Using this measure there is no statistical difference between the levels of NPD success in the four groups of firms studied.

The key findings on *Performance Measures* are summarised below:

1. Large Irish electronics firms launch many new products each year but generate a relatively low percentage of their revenue from products at the introduction stage. This is probably due to the incremental nature of the products that they develop.
2. Small Irish firms launch few new products each year but the contribution from products at the introduction stage is comparable to that in small UK firms. This suggests a heavy reliance on each new product launched.

8.2.3 Development Process Formality

The development process was examined at a company and at a project level. The project level measures examined the time spent at different NPD process stages and also the link between proficiency at the different stages and new product success. The company level measures concentrated on the formality of the process.

No statistically significant differences were found between the 6 groups of firms examined on any of the parameters measured. However, examining the mean scores of these parameters shows that small firms in both countries are:

- Less proficient at recording results,
- Less likely to hold post launch meetings,
- More likely to have extensive discussions outside formal meetings, and
- Weaker at organising product history files.

In addition to the data presented in Chapter 7 questions relating to how proficiently test results were recorded and how frequently post launch meetings were held allowed respondents to respond that such activities were not performed. Table 8.6 summarises the percentage of firms from each group that omitted or did not perform particular tasks.

The following task is omitted:	All Irish Firms	Small Irish Firms	Large Irish Firms	All UK Firms	Small UK Firms	Large UK Firms
Recording of customer reactions to early designs [L7]	8.3%	13.6 %	0%	4%	20%	0%
Recording of safety test results [L8]	5.5%	4.5%	7%	12%	20%	10%
Recording prototype test results [L9]	0%	0%	0%	0%	0%	0%
Post launch meetings [L16a]	11%	13.6 %	7%	8%	0%	20%
A product history file is not available [L29]	19%	23%	14%	8%	20%	5%

Table 8.6 Ability to Learn – Percentage of Firms who Omit Tasks

As expected a picture emerges showing that small firms omit the recording of NPD results more often than large firms. There are two exceptions to this. Fewer small Irish electronics firms than large Irish electronics firms omit the recording of safety test results, however this is probably obligatory for most firms developing and selling electronics products. Also, fewer small UK electronics firms than large UK electronics firms omit post launch meetings; in this case the sample size for the small firms (N = 5) is too small to draw any conclusions. Small firms in all groups were also less likely to be able to product a product history file. Again this analysis shows that small firms record test results and keep product history files less frequently than large firms. This result is consistent with the findings of other researchers (Hoffman et al., 1998; Santarelli & Sterlacchini, 1990).

Finally it is interesting to examine whether or not there is a link between process formality and NPD success in any of the groups of firms studied. To do this composite measure can be devised as the mean of the data presented in Table 7.10 in Chapter 7. (Note that two measures, L21 and L29 must be inverted to develop a meaningful scale.). Table 8.7 shows the results of correlating process formality with commercial success as measured using A22 – the average commercial success of new product launched during the past 3 years. Process formality is not significantly linked with new product success. In other words the informality of the processes at small firms is not a disadvantage, the case studies presented in Chapter 9 will explore this issue further.

	All Irish Firms	Small Irish Firms	Large Irish Firms	All UK Firms	Small UK Firms	Large UK Firms
Process Formality – measured by L7, L8, L9, L16a, L21 and L29	-.216	.223	-.546[d]	.027	.866	.165

Note 1: Correlation coefficients statistically significant at: [a] $p \leq 0.01$, [b] $p \leq 0.05$, , [c] $p \leq 0.10$, [d] $p \leq 0.20$, two-tailed

Table 8.7 Correlation between Commercial Success and Process Formality

The key finding on *Development Process Formality* are summarised below:

1. NPD processes in small firms are less formal than those in large firms.
2. NPD process formality is not significantly linked with new product success.

8.2.4 Involvement of External Organisations

The data presented in Chapter 7 show that on average Irish firms interact only occasionally with external organisations with the exception of customers. Irish and UK firms both reported that they *often* involved customers in their NPD projects. Most of the data published about cooperation on NPD projects examine only whether or not firms involve external organisations not the frequency of the involvement (LaBahn et al. 1996; Veugelers, 1997; Eurostat, 2000; Love and Roper 2001). Table 8.8 summarises the percentage of firms by size and country who have had *any* involvement with external organisations on NPD projects.

	Univer-sities	Research Institutes	Compet-itors	Other Firms	Custom-ers	Suppliers
All Irish (N=36)	83%	61%	25%	75%	86%	58%
Small Irish (N=22)	77%	50%	18%	68%	77%	45%
Large Irish (N=14)	93%	79%	36%	86%	100%	79%
All UK (N=24)	92%	63%	58%	92%	96%	92%
Small UK (N=5)	80%	60%	60%	100%	100%	100%
Large UK (N=19)	95%	63%	58%	89%	95%	89%

Table 8.8 % of firms involving external organisations in NPD projects

The data presented in Table 8.8 are higher than those collected by the EU in the CIS2 survey in 1997/8 (European Commission, 2000) where only 19% of small manufacturing firms and 50% of large manufacturing firms were found to be involved in cooperation for NPD. However, it should be noted that the EU data presented in Chapter 1 are for all manufacturing firms including those not involved in NPD. Some of the trends seen in the EU data where large firms were found to be more likely to collaborate externally on R&D than small firms, are repeated above. An earlier study of small high-technology Irish firms (Hurst and O'Kelly, 1995) also found that small Irish firms had less interaction with external organisations than large Irish firms.

Within the Irish data, large firms reported significantly higher levels of interaction with customers and suppliers than small firms. EU data show that vertical cooperation, i.e. with customers and suppliers is the most common form of collaboration. Many authors have linked NPD-related cooperation with customers with success (MacPherson, 1997; Tidd et al. 1998; Sternberg, 2000; Freel, 2003). Yet Table 8.8 shows that 23% of small Irish electronics firms reported that they did not involved customers in NPD projects. In the UK only 5% of small electronics firms undertook NPD projects without involving customers.

Fifty-five percent of small Irish electronics firms reported that they did not cooperate with suppliers in developing new products. Because of their lower purchasing power and production volumes it is perhaps more difficult for small firms to get cooperation from their suppliers in developing new products, yet only 11% of small UK electronics firms did not involve suppliers in their NPD projects.

UK firms reported significantly higher levels of interaction with competitors and suppliers than Irish firms. This contradicts the EU data that showed Irish firms having higher levels of interaction than UK firms. When this point is examined further it is found that there are no significant differences in interaction with competitors and suppliers between large Irish and UK firms[4]; but significant differences were found between small Irish and UK firms[5]. This reinforces the finding that small Irish electronics firms operate in a particularly isolated manner when developing new products.

To investigate the importance of these findings it is possible to examine the correlation between external interaction on new product innovation projects and the commercial success rate of new products. These results are summarised in Table 8.9.

Some of these results are surprising. NPD cooperation does not emerge as a strong success factor – only cooperation with suppliers in large UK electronics firms shows a significant positive relationship with commercial success.

Cooperation with universities or research institutes was found to correlate *negatively* with success in all groups of firms in the study with the exception of small Irish firms. A small positive correlation was found between success in small firms and the involvement of research institutes, though Table 8.8 shows that only half of all small firms involved research institutes in their NPD efforts.

---

[4] The statistical significance of the differences between competitors $p = .095$; suppliers $p = .583$

[5] The statistical significance of the differences between competitors $p = .025$; suppliers $p = .039$

	Correlations[1] between Average Commercial Success [A22][2] and NPD Cooperation [A36 – A41]					
	Univers-ities	**Research Institutes**	**Competitors**	**Other Firms**	**Customers**	**Suppliers**
All Irish (N=36)	-.279[c]	-.191	-.083	.158	.035	-.122
Small Irish (N=22)	-.086	.044	.138	.063	.099	-.261
Large Irish (N=14)	-.621[b]	-.598[b]	-.304	.332	-.118	.110
All UK (N=24)	-.584[a]	-.353[c]	-.046	.292[d]	.197	.440[b]
Small UK (N=5)	-.287	-.177	.000	-	.333	.395
Large UK (N=19)	-.624[a]	-.371[d]	-.090	.307	.195	.509[b]

Note 1: Correlation coefficients statistically significant at: [a] $p \leq 0.01$, [b] $p \leq 0.05$, , [c] $p \leq 0.10$, [d] $p \leq 0.20$, two-tailed
Note 2: The scale used for A22 (1 – Far above expectations to 5 – Far below expectations) has been inverted so that a positive correlation coefficient indicates a positive relationship with success.

Table 8.9 Average Commercial Success vs. NPD Cooperation

The negative impact of involving universities in NPD projects could be due to the inability of firms to work with universities and absorb the knowledge available therein, or alternatively (perhaps additionally) the lack of appropriate technology available in a suitable form from the universities. Difficulties in transferring technology from universities and government-funded research to industry have been documented (Spann et al. 1991; Hoffman et al. 1998). The work of Rothwell and Dodgson (1991) would suggest that the solution to this lies primarily within the firms own resources suggesting that Irish firm's need to improve their absorptive capacity (Cohen & Levinthal, 1989)

Unexpectedly, there is a negative association between the involvement of customers in NPD and NPD success in large Irish firms. The result is not significant and the sample size is quite small, however, there is little in the published literature that suggests that this should be the case.

Interaction with suppliers is found to be positively correlated with success in large Irish and all UK firms, but there is a negative association in small Irish firms. Only 45% of small Irish firms reported that they involved suppliers in their NPD projects and the data shown in Table 8.9 would suggest that it is not a strategy that increases NPD success rates in small firms. One possible explanation for this is that managing the involvement of suppliers in NPD projects is a skill that small Irish firms have not refined. The involvement of outsiders can complicate the NPD process to an extent that small firms cannot manage effectively – this view is supported by LaBahn et al. (1996) and will be examined in more detail in the case studies in Chapter 9.

Overall the results shown in Table 8.9 show that the involvement of external organisations in NPD does not have a strong and positive effect on NPD success in Irish or UK electronics firms; at best, the impact is neutral. This would support the findings of several authors (Spann & Souder, 1991; Hoffman et al., 1998; Freel, 2003), who found that internal resources are more important then external linkages in promoting successful NPD. LaBahn et al. (1996) also found that outsider involvement in NPD could have a negative effect, their study showed that it delayed product launch. Therefore; there is some clear support in the literature for the above findings.

Additionally, Table 8.9 shows that differences exist between the impact which interaction with external organisations on NPD success has in both small and large firms.

The key findings on *External Interaction* are summarised below:

- Irish electronics firms, in particular small Irish firms, have a low level of interaction with external organisations when developing new products.
- External interaction is not strongly associated with NPD success in the Irish or UK firms involved in this study.

8.2.5 Resources for NPD

Firms were asked about their spending on R&D as a percentage of sales and were also asked to allocate their product innovation spending between seven different categories: 1) basic research, 2) applied research, 3) process improvement, 4) NPD, 5) product quality improvement, 6) cost reduction and 7) marketing.

Significant differences were found in the level of R&D expenditure and R&D intensity between small and large firms in Ireland. Small firms spent over twice as much on product innovation and had almost twice as many R&D employees as large firms. These differences were also seen in the UK electronics firms though they were not found to be significant. In summary, R&D expenditure and intensity were found to be higher in small firms than in large firms. This is consistent with previous research that found that small firms have a higher R&D intensity than large firms (Acs and Audretsch 1988; Santarelli & Sterlacchini, 1990; Van Dijk et. al., 1997; Love and Roper, 2001). The difference in R&D intensity was significant in small vs. large Irish firms. This agrees with the results of a study by Andréosso-O'Callaghan (2000) who examined 85 firms in the Shannon region and found that 'no less than 16% of the labour force in the sampled indigenous electronics firms is made up of engineers, substantially more than the share of engineers in the foreign electronics industry'.

There was also a significant difference between the R&D intensity in Irish and UK electronics firms, UK firms reporting a higher intensity. When this result is explored in more detail, differences are found between the small firms in each country and also between the large firms in each country. Small Irish firms have a *higher* R&D intensity than small UK firms but large Irish firms have a *lower* R&D intensity than large UK firms. This is counter to expectations since large Irish firms are significantly smaller than large UK firms and therefore would be expected to have higher R&D intensity. In other words, the R&D intensity in large Irish firms is lower than expected.

Considering R&D expenditure with the performance measures discussed in Section 8.2.2 it can be seen that although small Irish firms invest *more* in NPD, in terms of number of employees (R&D intensity), they gain *less* from their results, in terms of percentage of turnover from new products. In other words, small firms are less effective in their NPD efforts than large firms. This result is supported by the findings of Santarelli & Sterlacchini (1990)

The data presented in Section 7.2.1 shows that small firms in both the UK and Ireland were concerned about the cost of NPD projects while large UK firms were concerned about funding NPD projects. Only large Irish firms were not concerned about financial resources for NPD projects. This is interesting particularly because R&D expenditure in large Irish firms was lower than in any other group of firms in the study. This supports the view that R&D is not a strategic activity for large Irish electronics firms – they are not investing heavily in NPD projects.

It is also interesting to examine R&D expenditure in relation to the number of new products launched per year. Table 8.10 summarises the R&D expenditure and number of new products launched in small and large, UK and Irish firms.

Small Irish firms spend a higher percentage of their sales revenue per new product launched than any other group of firms in the study, but in absolute terms, they have the least to spend. In both Ireland and the UK, small firms spend significantly less; in absolute terms, per new product launched.

Large Irish firms spend the lowest percentage of their sales revenue on each new product launched. Irish firms also spend a significantly lower percentage of revenue per new product launched than do UK firms. But when this result is further examined small Irish firms are not different from small UK firms (p = .803) while large Irish firms spend significantly less than large UK firms (p = .024). This would suggest that either (1) large Irish electronics firms are exceptionally effective in their NPD efforts or that (2) the products that they launch do not involve a significant amount of engineering expenditure i.e. they are mainly incremental product innovations. There is not much support in the data collected for the former view but the previous discussion would support the latter view that much of the new product development in large Irish electronics firms is incremental.

	Number of New Products [L36]	R&D expenditure as % of Sales [A33]	% R&D expenditure per new product	R&D ('000) expenditure per new product
All Irish (N=36)	3.25	11.5%	3.19%	IR£900
Small Irish (N=22)	2.0	13%	4.50%	IR£483
Large Irish (N=14)	8.0	6%	0.73%	IR£2,783
All UK (N=24)	4.0	11%	2.38%	ST£5,333
Small UK (N=5)	4.0	18%	2.14%	ST£750
Large UK (N=19)	4.0	9%	2.50%	ST£8,600
Comparison of Medians:				
Small vs. Large Irish	.033[a]	.012[a]	.008[a]	.001[a]
Small vs. Large UK	.803	.165	.767	.005[a]
Irish vs. UK	.652	.976	.971	.007[a]

[a] Statistically significant, $p \leq 0.05$, Mann Whitney U test

Table 8.10 R&D expenditure vs. number of new products per year. (Medians)

A relatively higher level of R&D expenditure as well as a higher R&D intensity is generally expected from small firms (Santarelli & Sterlacchini, 1990; Van Dijk et. al., 1997). Veugelers (1997) also found that while NPD spending increases with sales it does so less than proportionally. Therefore, as firms grow, (and their level of sales increases) their R&D expenditure, as a percentage of sales, could be expected to fall. Veugeler's study also found that R&D expenditure was lower in foreign owned companies; this was attributed to the centralisation of R&D and also due to the fact that many of the foreign-owned plants in the study were mainly involved in assembly. These factors also apply to the large Irish firms in this study and could account for their lower R&D expenditure.

The breakdown of product innovation spending (presented in Chapter 7) in Irish firms shows that the only significant difference between small and large firms was in the level of spending on process improvement. Large Irish electronics firms reported expenditure of 10% of their product innovation budget on process improvement while the corresponding amount for small firms is 3.5%. This result could be interpreted as a reflection of the increased importance of manufacturing within large firms when compared with small firms. However, it could also suggest that the product development activity carried out by the large manufacturing firms in the survey may be viewed more as a support to local manufacturing than as a strategic activity. The relatively low level of R&D spending in large Irish electronics firms also supports this argument.

Median expenditure on basic research in Irish and UK firms is zero. In Ireland the median expenditure on applied research is also zero while that in the UK is 5%. Irish firms spend significantly more on NPD and less on marketing new products than UK firms.

It is possible to examine the impact of the allocation for product innovation spending on commercial success. This is done in Table 8.10. Some of the results are very interesting; with the exception of large Irish firms (whose expenditure on Applied Research is very low), expenditure on basic or applied research correlates negatively with commercial success. It is useful to view this result in conjunction with the data presented in Table 8.9 which showed that interaction with universities was negatively correlated with success for all firms in the study. On the spectrum of product innovation-activity from basic research through to product development-the work undertaken by universities lies closer to the basic research end of the scale. The data presented in Tables 8.9 and 8.10-when taken together-reinforces the point that it is extremely difficult for firms of all sizes to successfully commercialise the results of basic research.

	Allocation of Product Innovation Expenditure [A23 – A29] vs. Average Commercial Success [A22][2]					
Activity	**All Irish Firms**	**Small Irish Firms**	**Large Irish Firms**	**All UK Firms**	**Small UK Firms**	**Large UK Firms**
Basic Research [A23]	-.005	-.195	.140	-.425[b]	-.395	-.427[c]
Applied Research [A24]	-.263[d]	-.354	-.216	-.549[a]	-.884[b]	-.500[b]
Process Improvement [A25]	.051	.035	.026	-.249	-.395	-.235
New Product Development [A26]	-.187	.019	-.314[d]	.380[c]	.648	.324[d]
Product Quality Improvement [A27]	.477[a]	.438[d]	.509[b]	.137	.412	.066
Cost Reduction [A28]	.333[b]	.327	.365[d]	.190	Note 3	.216
Marketing [A29]	.032	.170	-.046	.071	-.884[b]	.189
R&D expenditure as a % of sales	.184	.231	-.069	-.130	-.459	-.080

Note 1: Correlation coefficients statistically significant at: $^a\ p \le 0.01$, $^b\ p \le 0.05$, $^c\ p \le 0.10$, $^d\ p \le 0.20$, two-tailed

Note 2: The scale used for A22 (1 – Far above expectations to 5 – Far below expectations) has been inverted so that a positive correlation coefficient indicates a positive relationship with success.

Note 3: This item was constant (5%) for all five small UK firms.

Table 8.11 Product Innovation Expenditure vs. Commercial Success

It is very surprising that expenditure on NPD is negatively linked with success in large Irish firms. This statistic means the firms who spend more on NPD have a lower success rate with new products. Previous results have shown that large Irish electronics firms spend very little on product innovation and that they launch a large number of new products, mainly product variants, each year. This has been interpreted to show that these firms are more involved in incremental product innovation as opposed to radical product innovation. This view is supported by the positive correlations between expenditure on product quality improvement and cost reduction. Additionally, external interaction with universities and research institutes, both of whom would be more associated with basic research than product development, are negatively correlated with success in large Irish electronics firms. All of this data, when taken together builds a picture of large Irish electronics firms which are good at incremental product innovation – i.e. they launch many new products but spend relatively little on product innovation. Their expenditure on cost reduction and quality improvement is also linked with success; they are poor, however at more substantial or (radical) new product innovations – i.e. their collaborative efforts with universities and research institutes are not successful, and their increased expenditure on NPD is negatively linked with success.

The very weak positive correlation between NPD expenditure and commercial success in small Irish firms shows that the firms who spend a higher percentage of their product innovation expenditure on NPD do not necessarily achieve greater commercial success. Given that small Irish electronics firms already spend 6.5% of their turnover on launching each new product, -higher than any other group of firms-, this result could be interpreted to show that they are already investing enough in developing new products and greater success can be achieved by changing how they are using existing NPD resources, rather than by allocating more resources to this.

The key findings on *Resources for NPD* are summarised below:

- R&D expenditure and intensity is higher in small firms than in large firms; in absolute terms, however, small firms spend significantly less than do large firms per new product launched.
- Large Irish electronics firms spend a lower percentage of revenue per new product launched than do any other group of firms in the study. Additionally, they are not concerned about the cost of NPD. This suggests that these firms are primarily involved in manufacturing and supports the argument that they mainly undertake incremental product innovation.
- Expenditure on basic and applied research is very low in Irish electronics firms and is also not linked with NPD success. The INTERPROD data supports the view that it is extremely difficult for firms to successfully commercialise the results of basic research.
- Increased spending on NPD in small Irish electronics firms is not linked with increased NPD success; to increase success they need to change *how* they are spending on NPD, not how *much* they are spending on NPD.

8.2.6 Cycle Time Practices

Time-to-market is one of the key performance metrics in the NPD process; it is also a key concern for the firms involved in this study as was shown in Section 8.2.1. The INTERPROD survey measures time-to-market as the average time from project initiation to commercialisation. For most of the firms studied this was between 6 months and 3 years. Previous studies of small firms suggest that small firms have the advantage of 'much shorter times to market' (Walsh, 2002). This was found to be the case within the UK firms where time-to-market was statistically significantly shorter in small firms than in large firms. The mean time to commercialise a new product was also lower in small Irish firms than in large Irish firms, although not significantly so.

Time-to-market was also found to be significantly shorter in Irish than in UK firms. Examining this result in more detail shows that small Irish firms are not significantly different from small UK firms (p = .110) but time-to-market is significantly shorter in large Irish firms than in large UK firms (p = .012). One explanation for this difference could be that large Irish firms are involved in 'smaller' NPD projects and the data presented in the previous section bears this out.

The INTERPROD cycle time questionnaire (INT-F) examined issues such as the degree to which users and suppliers were involved in NPD, the flexibility of personnel, products and the consideration for manufacturing during design. Within the Irish sample, large firms were found to be significantly more likely to consider manufacturing details during design and were also more likely to partner with their suppliers. This supports the previously discussed findings on external interaction where small Irish firms were found to interact less frequently with suppliers; it also suggests that large firms are more concerned about manufacturing than small firms.

Firms were asked about techniques that they had used to reduce NPD cycle times. Table 8.12 summarises these findings.

Challenge	Firms listing this technique amongst its top 3					
	All Irish	All UK	Small Irish	Small UK	Large Irish	Large UK
Anticipating customer's future needs	X	X	X	X	X	X
Attention to manufacturability during design	X		X	X	X	
Close collaboration between R&D and manufacturing.	X		X		X	
Setting clear project objectives		X	X	X		X
Setting clear time goals		X		X	X	X
Improved project planning		X		X	X	X
Adoption of ISO9000					X	X
Establishing clear priorities					X	X

Table 8.12 Commonly Used Cycle Time Reduction Techniques

All the large firms cited *Adoption of ISO9000 standards* as a commonly used technique for reducing cycle times. Small firms used this less frequently. The use of ISO9000 would be linked with the formalisation of processes. Small firms have already been seen to have less formal processes; therefore, this result does not come as a surprise. There is also support from the literature for the fact that small firms are less formal about their NPD efforts than large firms (Santarelli & Sterlacchini, 1990). While the adoption of ISO9000 was listed as a commonly-used technique, it was not found to be useful in reducing cycle times, see Table 8.13.

Small Irish firms used fewer project management techniques to reduce cycle time than other firms – large Irish and UK firms commonly used improved project planning and setting clear time goals and priorities to reduce cycle times; none of these were among the top three techniques used by small Irish firms.

Table 8.13 summarises the techniques that were found to be most *effective* in reducing cycle times in each group of firms.

Challenge	**Firms listing this technique amongst its top 3**					
	All Irish	**All UK**	**Small Irish**	**Small UK**	**Large Irish**	**Large UK**
CAD/CAM	X	X	X		X	X
Improved project planning	X		X			X
Close collaboration between R&D and manufacturing.	X			X	X	
Anticipating customers future needs		X		X		X
Greater team responsibility (empowerment)		X		X		X
Customer inputs at the design stage	X				X	X
Attention to manufacturability during design		X			X	X
Establishing clear priorities	X		X	X		
Setting clear project objectives		X	X	X		

Table 8.13 Effective Cycle Time Reduction Techniques

Overall, the most effective cycle time reduction technique is the use of CAD/CAM. This is consistent with the findings of previous research (Dröge et al. 2000). Both large firms found that *Customer inputs at the design stage* and *Attention to manufacturability during design* were effective techniques for reducing NPD times. There is some support for this in the literature. Both of these techniques would be facilitated through the use of cross-functional teams; this is identified by many authors as a method for reducing time-to-market (Eisenhardt & Tabrizi, 1995; Zirger & Hartley, 1996; Griffin, 1997). The fact that neither of these techniques was particularly effective in the small firms is of interest.

It can possibly be explained by the fact that the techniques which have been proved useful in reducing cycle times in small firms relate to basic project planning and control (e.g. improved project planning, establishing clear objectives and priorities). The small firms in this study, particularly the small Irish firms, do not appear to be using integrated NPD process.

Attention to manufacturability during design and *Close collaboration between R&D and manufacturing* are both commonly used by small firms, see Table 8.12, but neither is found to be effective. This could be an example of small firms adopting techniques that are generally perceived as 'best practice', but simply do not apply to them.

By contrast, both groups of small firms listed *Setting clear project objectives* and *Establishing clear priorities* amongst their most effective techniques. There is support in the literature for the use of time as a project goal being linked with shorter NPD time (Zirger & Hartley, 1996). However, small Irish firms do not frequently set clear time goals, establish clear priorities or plan projects, see Table 8.12. This suggests that small Irish firms could reduce their NPD time by developing clearer project definitions (including objectives, goals and priorities). Additionally, small Irish firms do not frequently use improved project planning to reduce NPD cycle-times although this has been found to be an effective technique for them. The effectiveness of project planning as a cycle-time reduction technique is contrary to previous research (Zirger & Hartley, 1996; Datar et al. 1997). Both of these studies were based on large firms who could be assumed to have reasonable formal NPD processes, however; Datar's firms consisted of 3 international manufacturers of electronics components while Zirger & Hartley surveyed 44 firms that averaged $260 million in sales. It is possible that for most firms increasing project planning beyond a certain level does not improve NPD times but firms who do not do any project planning could benefit by doing some, however.

Finally, it is useful to examine the extent to which NPD cycle time has an impact on new product success. Table 8.14 shows the results of correlating the average time from project initiation to commercialisation [A34] with the average NPD success rate [A22]. The results show that the only group of firms for which cycle time is linked with success is large UK firms. Also, the positive link between NPD project time and success in large Irish and small UK firms suggests that for these firms rushing NPD projects can result in lower success rates.

	All Irish Firms	Small Irish Firms	Large Irish Firms	All UK Firms	Small UK Firms	Large UK Firms
Average time from project initiation to commercialisation [A34]	-.037	-.134	.112	-.309	.354	-.484[b]

Note 1: Correlation coefficients statistically significant at: $^{a}\ p \leq 0.01$, $^{b}\ p \leq 0.05$, , $^{c}\ p \leq 0.10$, $^{d}\ p \leq 0.20$, two-tailed

Table 8.14 Correlation between Commercial Success and NPD Cycle Time

In summary, the results for cycle-time practices have highlighted several differences between the management of NPD in small and large firms. These are summarised below.

The key findings on *Cycle Time Practices* are summarised below:

1. Large Irish firms have significantly shorter time-to-market than large UK firms. The most likely explanation for this is that the Irish firms are involved in smaller, incremental, projects.
2. Small Irish firms use fewer project management techniques than any other group of firms. Yet, the small Irish firms who have used project management techniques to reduce NPD cycle times have found them to be very effective.
3. Fewer small firms than large firms adopted ISO9000 standards; this supports the theory that small firms have a less formal approach to NPD.
4. Considering manufacturing during design was effective in reducing cycle time for large Irish firms but not for small Irish firms. This again emphasises the relative importance of manufacturing to large Irish firms.
5. Shorter NPD cycle times are not always linked with increased success.

8.2.7 Summary of Company-Level Results

Table 8.15 summarises all of the company level results. The key findings from each of the seven areas of company level results allow a picture to be built up of small and large electronics firms in Ireland and the UK. Since the focus of this study is on small Irish electronics firms this group of firms will be discussed in most detail. However, some interesting results that have emerged concerning large Irish electronics firms will also be examined. It is important to look at both small and large UK firms in order to place the findings about the Irish firms in context. What follows here is a description of small and large Irish and UK electronics firms based on the information collected by the company level INTERPROD questionnaires.

Small Irish Electronics Firms

The small Irish firms in this study are all heavily committed to NPD. They invest significantly in R&D; this is seen both in the R&D intensity in small firms and also in the percentage of their turnover that they invest in developing new products. In term of inputs to the NPD process, small firms perform better than large firms. Also, small Irish firms are confident about their ability to develop and commercialise new products and they judge a high percentage of their new products to be successful. However, small firms do not perform as well when the outputs of the NPD process are measured. Small firms generate less revenue from new products and from products at an introduction stage than do large firms - they launch fewer new products each year than any other group of firms in the study. At a broad level, the situation in small Irish electronics firms can be summarised by noting that they invest heavily in developing new products but are not getting the same return for their investment as large firms.

It is important then to try to understand why this is the case; some answers are provided in the company level data collected by INTERPROD. Small firms are still concerned with NPD operational issues. This is shown in the concerns that they express – improving quality, reducing failure; and also in the techniques that they use, or rather fail to use – small firms are not using project management techniques. They are also found to have less formal NPD processes than large firms. Small firms do not interact extensively with external organisations and have not yet developed the skill to benefit from supplier involvement in developing new products.

What works for small firms developing new products? Spending more money has been shown not to improve their success; however, improved project definition and planning are useful in getting new products to market faster.

Large Irish Electronics Firms

One of the most striking features of large Irish electronics firms is how little they invest in each new product launched – their investment in NPD is relatively low yet they launch more new products per year than any other group of firms in the study. Their R&D intensity is also the lowest in the sample. In terms of inputs to the NPD process large Irish electronics firms perform very poorly. When the NPD output of large Irish firms is examined they are also found to be poor performers – they generate significantly less revenue from products at the introduction stage than large UK firms. Large Irish electronics firms also have the shortest time-to-market of any of the firms in the study. All of this suggests that large Irish electronics firms are mainly involved in incremental NPD.

Manufacturing also emerges as a key concern for large Irish electronics firms. Manufacturability is an important design consideration for them and they spend significantly more on process improvement than small Irish firms. Expenditure on quality improvement and cost reduction are both linked with success in large Irish firms. Overall, the results relating to large Irish electronics firms suggest that they are primarily manufacturing facilities that are involved in some incremental R&D.

Small UK Electronics Firms

Many of the characteristics of small Irish electronics firms are seen in small UK electronics firms. They also invest a much higher percentage of their turnover on new products than large UK firms; in fact small UK firms spend more on product innovation than any other group of firms in the study. Small UK firms are not as isolated in their NPD efforts as small Irish firms; all the small UK firms reported that they interacted with suppliers and customers in developing new products. Small UK firms share many of the same concerns as small Irish firms – i.e. NPD costs, product quality and failure rates. Small UK electronics firms differ from small Irish firms in their use of project planning and definition. (It needs to be stressed that there were only 5 small UK electronics firms involved in the study; this limits the conclusions that can be drawn from this group of firms.)

Large UK Electronics Firms

To some extent large UK electronics firms are used as the benchmark for the other firms in this study. There were few unexpected results from this group of firms. They invest a reasonably high percentage of turnover on developing new products – however, based on the EU data (see Chapter 1), this would be expected from large firms in this sector. They also generate a high percentage of their revenue from new products or products at the introduction stage. Their interaction with external organisations is on a par with that of other groups of firms. In common with the other groups of firms, interaction with universities and research institutes -and also basic and applied research- is negatively correlated with success in large UK electronics firms.

1. The key findings on *NPD Context* are summarised below:

1.1 The UK firms involved in this study were significantly larger than the Irish firms in terms of both size and turnover.

1.2 Large firms are concerned about future opportunities - new product ideas, new markets, and improved time-to-market. Small firms are still concerned about operational issues - controlling costs, improving quality and reducing failure.

2. The key findings on *Performance Measures* are summarised below:

2.1 Large Irish electronics firms launch many new products each year but generate a relatively low percentage of their revenue from products at the introduction stage. This is probably due to the incremental nature of the products that they develop.

2.2 Small Irish firms launch few new products each year but the contribution from products at the introduction stage is comparable to that in small UK firms. This suggests a heavy reliance on each new product launched.

3. The key findings on *Development Process Formality* are summarised below:

3.1 NPD processes in small firms are less formal than those in large firms.

3.2 NPD process formality is not significantly linked with new product success.

4. The key findings on *External Interaction* are summarised below:

4.1 Irish electronics firms, and in particular small Irish firms, have a low level of interaction with external organisations when developing new products.

4.2 External interaction is not strongly associated with NPD success in the Irish or UK firms involved in this study.

5. The key findings on *Resources for NPD* are summarised below:	
5.1	R&D expenditure and intensity are higher in small firms than in large firms; but in absolute terms small firms spend significantly less than large firms per new product launched, however.
5.2	Large Irish electronics firms spend a lower percentage of revenue per new product launched than any other group of firms in the study. Additionally, they are not concerned about the cost of NPD. This suggests that these firms are primarily involved in manufacturing and supports the argument that they mainly undertake incremental NPD.
5.3	Expenditure on basic and applied research is very low in Irish electronics firms and is also not linked with NPD success. The INTERPROD data supports the view that it is extremely difficult for firms to successfully commercialise the results of basic research.
5. 4	Increased spending on NPD in small Irish electronics firms is not linked with increased NPD success; to increase success they need to change *how* they are spending on NPD, not how *much* they are spending on NPD.
6. The key findings on *Cycle Time Practices* are summarised below:	
6.1	Large Irish firms have significantly shorter time-to-market than large UK firms. The most likely explanation for this is that the Irish firms are involved in smaller, incremental, projects.
6.2	Small Irish firms use fewer project management techniques than any other group of firms. Yet the small Irish firms who have used project management techniques to reduce NPD cycle times have found them to be very effective.
6.3	Fewer small firms than large firms adopted ISO9000 standards; this supports the theory that small firms have a less formal approach to NPD.
6.4	Considering manufacturing during design was effective in reducing cycle time for large Irish firms but not for small Irish firms. This again emphasises the relative importance of manufacturing for large Irish firms.
6.5	Shorter NPD cycle times are not always linked with increased success.

Table 8.15 Summary of Company Level Results.

8.3 Project Level Results

The analysis and discussion of the project level results is divided into eight sections:

1. R&D/Marketing Integration – Section 8.3.1
2. Project Management Skills – Section 8.3.2,
3. Top Management Support – Section 8.3.3,
4. Development Process Proficiency – Section 8.3.4,

5. New Product Characteristics – Section 8.3.5,
6. Market Characteristics – Section 8.3.6,
7. Marketing and Technological Skills – Section 8.3.7,
8. Marketing and Technological Synergy – Section 8.3.8.

8.3.1 R&D/Marketing Integration

As expected, (Alder et al., 1989; Griffin & Hauser, 1996; Leenders & Wierenga, 2002) the level of contact and information flow between commercial and technical entities correlated significantly with project success in all Irish firms and all UK firms. However, when the samples are split into small and large firms a significant correlation is seen in the small firms but not in the large firms. The level of participation by commercial entities in problem solving (concerning new products) shows a similar trend – the correlation with success in small firms is much stronger than that in large firms. This is the opposite of what the literature reviewed in Chapter 3 would suggest.

A strong correlation between any management factor and project success suggests that this factor was managed differently in successful and failed projects. Therefore, the lack of a relationship between commercial/technical communications in large electronics firms could suggest that this factor is consistently managed on successful and unsuccessful projects in large firms. The work of Leenders & Wierenga (2002) would support the view that the R&D-marketing interface is well managed in large firms through the use of formal processes, cross-functional teams and ICT.

However, the presence of a relationship between commercial/technical communications and success in small firm suggests that not all NPD projects in small firms have good communications between R&D and marketing entities. This contradicts conventional theory about small firms having 'open communication channels' (Berry & Taggart, 1998) and 'efficient internal communication' (Rothwell & Dodgson, 1991a).

One possible explanation for this is that efficient communications do exist in very small firms but that as a firm grows, interdepartmental communications become more difficult – this is a view supported by Griffin & Hauser (1996). Large firms tend to have more formal processes and to use other mechanisms (Leenders & Wierenga, 2002) to facilitate R&D-marketing integration. However, firms that have grown too large for the open communications associated with very small firms and that have not yet put in place the more formal processes associated with larger firms may find themselves in a sort of *adolescent* stage where effective R&D-marketing communication happens on some but not all NPD projects and is a differentiator between new product success and failure.

This theory can be tested by looking at the correlation between success and these two factors in very small firms (fewer than 20 employees) and small firms (between 20 and 100 employees) within the Irish firms. Table 8.16 shows that there is support for the view that as firms grow they go through a phase where they have lost some of the benefits of being very small - flexibility, interactive management, efficient communications, (Rothwell & Dodgson, 1991) and have not yet gained the benefits of being large - formal procedures, well established lines of communication (Leenders & Wierenga, 2002)

	Very Small Firms (<20 employees) n = 20	**Small Firms (20< employees <100) n = 27**
Level of information flow between commercial and technical entities	.279	.653[a]

Note: Correlation coefficients statistically significant at: [a] $p \leq 0.01$, [b] $p \leq 0.05$, two-tailed

Table 8.16 Information Flow vs. Success in small and very small firms.

The key findings on *R&D/Marketing Integration* are summarised below:

1. Larger, small firms (20 – 100 employees) appear to have difficulty in managing commercial and technical communications. This was not an issue for very small (fewer than 20 employees) firms or large firms.

8.3.2 Project Management Skills

One of the most striking differences between the organisational factors in UK and Irish firms is the degree to which Project Management Skills correlate with success. All the project manager characteristics listed in Table 7.18 correlated with success in all UK firms. In all Irish firms, only two project manager characteristics, participative style and authority from top management, correlated with success to any significant degree, all other relationships were weak.

Taking these results at face value it could be assumed that project managers in Irish firms are more skilled than those in UK firms. Using the same logic applied in the discussion of R&D-marketing communications, this could be interpreted to show that project management is consistently at a high level in Irish electronics firms but is managed less consistently in UK electronics firms. But, if the company level findings, which showed that Irish firms are not using project management techniques, are taken into consideration it more likely that the role of project manager is not well defined in many Irish firms and therefore the project managers skills are less likely to have an impact on project success.

The absence of any significant relationship between project planning and control in Irish firms suggests that Irish firms are not engaging in and benefiting from project management methods in the development of new products. This is in line with the finding that project managers skills do not correlate with project success in Irish firms.

The key findings on *Project Management Skills* are summarised below:

1. Project Management Skills do not correlate with project success in Irish firms - results suggest that the project manager's role is not well defied in Irish firms.
2. Project planning and control is not linked with new product success in Irish firms.

8.3.3 Top Management Support

Another factor that differentiates small electronics firms from large electronics firms is the impact of top management support on project success. The results in Chapter 7 show that top management support is significantly correlated with success in small firms but not in large firms. It is normally assumed that new product projects are not initiated within small firms without the support of top management (Maidique, 1980), while the support of top management is frequently cited as a success factor in studies of large firms (Cooper & Kleinschmidt, 1995; Swink, 2000). The data in Table 8.17 indicate that the support of top management does not differentiate between successful and unsuccessful projects in large firms. But in small firms projects that have the support of top management are more likely to be successful. Again, this finding contradicts the literature reviewed in Chapter 3.

The concept of an adolescent stage described in Section 8.3.1 may also provide an explanation for the impact of top management support on project success in small and large firms. As a firm grows it is no longer possible for top management to closely monitor all activities within the firm. Until firms put in place formal procedures for project identification, selection and resourcing the support of top management for a project can make the difference between success and failure.

	Very Small Firms (<20 employees) n = 20	**Small Firms (20< employees <100) n = 27**
Top management support	.311	.404[b]

Note: Correlation coefficients statistically significant at: [a] $p \leq 0.01$, [b] $p \leq 0.05$, two-tailed

Table 8.17 Top Management Support vs. Success in small and very small firms.

The weak negative correlation between top management support and project success in large UK firms is possibly an example of top management 'interference' (Swink, 2000). In large high-technology firms it is difficult for top management to understand all the issues involved in developing new products. Increased attention from top management is as likely to be a hindrance as a help.

The key findings on *Top Management Support* are summarised below:

1. Larger, small firms (20 – 100 employees) appear to have difficulty in managing top management support. This was not an issue for very small (fewer than 20 employees) firms or large firms.

8.3.4 Development Process Proficiency

Firms were asked about how they allocated time between the different stages of new product development. They were also asked to rate their performance at each of the eight stages presented: 1) exploratory, 2) concept development, 3) prototype development, 4) prototype testing, 5) market development, 6) manufacturing start-up, 7) marketing start-up, and 8) technical service. These data were then correlated with project success levels to allow some conclusions to be drawn about the practices linked to NPD success.

The first set of graphs presented in Section 7.3.4 gave an overview of the time spent by each group of firms at each of the eight NPD stages. These data were not analysed based on project success or failure but simply examined for differences between the four groups of firms. Within the Irish firms no significant differences were found between how small and large firms allocate NPD time.

The time spent at the different stages of NPD was summarised into three types of activities; pre-development, development and product introduction, this was then examined in small and large and on successful and unsuccessful projects – Table 7.20. When all the firms were analysed together it emerges that significantly more time is spent on pre-development activities and less on product introduction for successful new products. There is much support for the finding that the most critical steps in the NPD process are those included in predevelopment activities (Cooper & Kleinschmidt, 1986, 1987; Parry & Song, 1994, Millson & Wilemon, 1998) therefore when more time is allocated to predevelopment activities a new product has a greater probability of success. There are also several reasons why it should take longer to introduce less successful new products; a lower market demand would make a new product a lower priority, problems during the development stage could make a product more difficult to introduce to manufacturing, problems in the field could mean that more time is spent servicing a less successful new product.

In large firms more time is spent at pre-development activities in successful than in unsuccessful new products – this reconfirms the above finding. However, in small firms significantly more time is allocated to *product development* in unsuccessful projects. This highlights product development as a critical activity in small firms and could perhaps be used as an early warning of project failure for small firms.

Results were also presented in Chapter 7 showing the relationships between proficiency at the different stages of NPD and project success, these data were shown in Table 7.21. Many significant correlations were found between proficiency in the different NPD stages and project success, for simplicity Table 8.18 shows the two most significant activities for small and large, UK and Irish firms.

Small Irish Firms	**Large Irish Firms**
Prototype Development	Manufacturing Start-up
Market Development	Marketing Start-up
Small UK Firms	**Large UK Firms**
Manufacturing Start-up	Marketing Start-up
Marketing Start-up	Prototype Development

Table 8.18 Most critical NPD activities.

Success in small Irish firms is correlated significantly with their ability to develop new products and to develop markets for these new products. The importance of these two factors is not surprising. A study by Cooper & Kleinschmidt (1986) linking the proficiency of development activities with new product success found that the most important activities were (1) initial screening and marketing research, (2) business analysis (not examined in this study) and (3) product development. Previous research would also support the view that small firms involved in product innovation and are most concerned about technical issues i.e. product development (Abernathy & Utterback, 1978; Maidique, 1980). Large UK firms are similar to small Irish firms and again, the importance of product development and marketing is supported by the literature.

Success in small UK firms and large Irish firms is linked with manufacturing and marketing start-up, both of these are classified as product introduction activities. This supports the focus on manufacturing in large Irish electronics firms that was identified in the company level data. The importance of a proficient market launch has also been identified in previous publications (Debruyne et al. 2002; Di Benedetto, 1999).

Two previous studies that differentiated between radical and incremental new product projects (Song & Montoya-Weiss, 1998; Rochford & Rudelius, 1997) concluded that market analyses/development was a more critical activity for more radically new products. Viewing the Irish and UK INTERPROD results in this light suggests that small Irish firms are involved in more innovative new product projects than large Irish firms.

In summary the key findings on *Development Process Proficiency* are:

1. Small firms spend significantly longer at product development activities in unsuccessful projects than in successful projects. The length of time spent at development activities could be used as an early warning system for new product failure in small firms.
2. For all the firms in the study, more time is spent at predevelopment activities and less time at product introduction activities on successful NPD projects than on unsuccessful NPD projects.
3. The relative importance of the different NPD activities to small Irish firms suggests that they are involved in more innovative projects than large Irish firms.

8.3.5 New Product Characteristics

Only two new product characteristics (1) products that are cost effective and (2) products that offer benefits and solutions to customers, were linked (weakly) with success in all Irish and all UK firms. This is consistent with previous research findings (Cooper, 1994).

The only product characteristic that correlated significantly with success in small Irish firms was product compatibility; this is consistent with a previous study (Yap and Souder, 1994). A positive correlation was also found between compatibility and success in small firms in the UK but the relationship was negative for both Irish and UK large firms. Using product reliability as a planned selling point was also positively correlated with success in each group of small firms (though only significantly in small UK firms), while it was negatively correlated with success in UK and Irish electronics firms. This supports the findings from the company level results that the reliability of new products is still an issue for small firms but not for large firms.

In summary the key findings on *New Product Characteristics* are:

1. Product reliability and compatibility are positively linked with success in small but not large firms. This reconfirms company level findings that small firms are still concerned with basic operational issues.

8.3.6 Market Characteristics

Firms involved in the study were asked a range of questions about customer and market characteristics. Their responses to these questions were then correlated with the level of new product success. Significant results were presented in Section 7.3.6.

There are only two results that are consistent in both Irish and UK firms.

- Markets that are growing quickly correlate significantly with new product success in large but not small firms.
- There is a negative relationship between launching new products into mature markets in large firms but not in small firms.

A possible explanation for these results is that quickly growing markets are seen as opportunities by many firms and therefore tend to be competitive; it is difficult for small firms, with limited resources, to compete in such markets. Also, there is evidence that small firms avoid direct competition with large firms (O'Gorman, 2000) – this would be likely to happen in quickly growing markets. On the other hand mature markets tend to be seen as less attractive by larger firms but can provide niche opportunities for small firms (Chaganti, 1987). Therefore the above results can be interpreted as reinforcing conventional wisdom about small firms avoiding direct competition with large firms and seeking market gaps or niches (Hoffman et al. 1998).

The customer characteristics that were most strongly linked with new product success in small Irish firms were:

- Customers who had a great need for the product.
- Customers who were eager to take on new products.
- Considering customers' problems with previous problems when designing new products.

A customer having a great need for a particular new product suggests that this type of product is not readily available. The fact that success in NPD in small Irish firms is linked to this customer characteristic suggests that small Irish firms are willing to develop products specifically to meet customer requirements/needs. In addition, success in small Irish firms was linked with customers who were eager to take on new products. These findings could be summarised by saying that success in small Irish firms is linked with developing innovative products for specific/niche customer needs.

None of the customer characteristics correlated significantly with success in large Irish or UK firms. This suggests that large firms do not work closely with individual customers in developing new products. It is possible that they target new products at markets rather than at specific customers.

Having a good relationship with customers was linked with new product success in UK firms (weakly in the case of large firms) as was developing products for customers who were dissatisfied with existing products. Having good relationships with customers is more important for UK firms than for Irish firms. This could be related to the high level of export in Irish firms. Physically, Irish firms are not close to their customers; they seem to compete through the provision of innovative, niche products rather than good customer relationships.

Finally, a negative relationship was revealed between new product success and unveiling product concepts too late in all firms except large Irish firms. This would support the view that large Irish firms are not involved in really new products for which time-to-market is critical (Vessy, 1991; Griffin, 1997).

In summary the key findings on *Market Characteristics* are:

1. Results relating to the impact of market growth and maturity on the success of products launched by small firms suggest that small firms avoid direct competition with large firms and that they seek market gaps and niches.
2. Success in small Irish firms is linked with developing innovative new products for specific/niche customer needs while success in small UK firms is linked with good customer relationships.
3. Two trends are indicated by the relationships between customer characteristics and new product success; (1) UK firms are closer to their customers than Irish firms, and (2) Large firms develop new products for markets rather than specific customers.

8.3.7 Marketing and Technological Skills

The relationships between technical skills and new product success were not as clear or as strong as those between marketing skills and success - particularly in the Irish firms. This is a surprising result especially for technology-based firms where it could be expected that technical skills should be critical (Song & Montoya-Weiss, 1998).

Within the small Irish firms, none of the engineering/technical skills were significantly ($p \leq 0.05$) correlated with new product success. In large Irish firms only the level of manufacturing skills was significantly correlated with project success, another indication of the importance of manufacturing to large Irish electronics firms.

Small UK firms were similar to small Irish firms - none of the engineering/technical skills were correlated with new product success. However in large UK firms, engineering and technical skills were found to correlate with success. This is as expected and supports the earlier findings relating to the link between proficiency at product development and new product success in large UK firms (Cooper & Kleinschmidt, 1987; Song & Montoya-Weiss, 1998).

In Ireland and the UK the relationship between resources availability and success is higher in small than in large firms – this supports earlier findings that resourcing NPD is a concern for small firms.

In summary the key findings on *Marketing and Technological Skills* are:

1. Marketing skills are found to be more critical to new product success than technological skills.
2. Resource availability is a more critical issue for small than large firms.

8.3.8 Marketing and Technological Synergy

Three of the variables in Table 7.25 examine the degree of fit between a firm's existing markets and products and the new product being launched. All three variables are found to be significantly correlated with success in small Irish and large UK firms. A strong correlation indicates that firms are more likely to be successful with new products that are closely aligned with their existing markets and product lines. Cooper & Kleinschmidt (1995) also found that market synergy was linked with success. To explore this result in more detail Table 8.19 shows the mean results for these three variables.

Factor[1]	Small Irish Firms	Large Irish Firms	Small UK Firms	Large UK Firms
There was close fit between existing markets and the market for this product. [C52]	3.46 (1.19)	4.04 (1.08)	3.40 (1.38)	3.62 (1.74)
There was close fit between the existing product line and this product. [C53]	3.41 (1.17)	3.85 (1.05)	3.40 (1.38)	3.59 (1.24)
There was a close fit between the firms marketing skills and the needs of this project [C54]	3.20 (1.18)	3.42 (1.33)	2.60 (1.38)	3.44 (1.83)

Note 1: 1 – strongly disagree to 5 – strongly disagree

Table 8.19 Degree of Fit, Mean (Standard Deviation)

The new products launched by large Irish firms are closer to existing product lines and markets than those launched by any other group of firms. Also, the lower standard deviations for large Irish firms show that this finding is consistent within this group of firms. In other words, large Irish firms stay very close to their existing product lines and markets in their NPD activities – again evidence that they are involved mainly in incremental NPD. Interestingly small firms, particularly small UK firms, develop new products that are farther from their existing markets and product lines than large firms.

In summary the key findings on *Marketing and Technological Synergy* are:

1. Large Irish firms develop new products that are close to their existing product lines and markets, i.e. incremental products
2. Small Irish and large UK firms are more successful with new products that are close to their existing product lines and markets – market synergy is a success factor.

8.3.9 Methods of Organising NPD

The most popular methods of organising NPD projects in UK and Irish firms were using NPD departments and/or NPD managers. Previous research would suggest that organising NPD under product/marketing managers (Barczak, 1995) or NPD managers (Lee et al., 2000) are the most popular methods of organising NPD in US firms. Interestingly NPD departments were not common in the US but were used extensively in Korea (Lee et al., 2000). In the US project teams headed by technical people are linked with success. Within the Irish and UK data there were no statistically significant differences between new product success and the use of different methods for organising NPD.

Finally, it is interesting to note that in Irish firms there is no link between the formality of NPD procedures and new product success. This supports the earlier finding, Section 8.2.3. Table 8.20 shows the degree to which formal procedures were used in small and large Irish and UK firms – large Irish firm use more formal procedures than any other group of firms, and increased formality correlated negatively with success for this group of firms. For the UK firms there is a weak positive link between procedure formality and new product success.

Factor[1]	**Small Irish Firms**	**Large Irish Firms**	**Small UK Firms**	**Large UK Firms**
The procedure for carrying out the tasks on this project was formal [B13]	3.22	4.04	3.10	3.55

Note 1: 1 – strongly disagree to 5 – strongly disagree

Table 8.20 Formality of NPD procedures, Means

In summary the key findings on *Methods of Organising NPD* are:

1. Increased procedural formality is not significantly linked with success in any of the firms studied.

8.3.10 Summary of Project Level Results

Table 8.21 summarises all the project level results. The same process that was used to summarise the company level results will be applied to the project level results. A summary of the findings relating to each group of firms will be presented, again the emphasis will be on the small Irish firms.

Small Irish Electronics Firms

From an organisational perspective small Irish firms cannot be treated as a homogeneous group. Very small firms (fewer than 20 employees) exhibit the traits commonly associated with small firms; an organic structure with effective internal communication and a high level of top management involvement (Rothwell & Dodgson, 1991). But larger small firms (20 to 100 employees) appear to outgrow this organisational model and experience some of the problems often associated with large firms.

The role of the project manager is not well defined within small Irish firms and there is evidence that these firms are not seriously involved in planning and controlling NPD projects. New product success in small (but not large) Irish firms is linked to product reliability and compatibility, and the availability of resources, in other words small firms are still struggling with operational NPD issues.

From a marketing perspective the project level data show that small Irish firms are successful at developing innovative products that solve specific customer needs and that they have most success competing within market niches – not head-to-head with large firms. Finally, product development is a key activity for small Irish firms. There is a closer link between engineering skills and new product success in small than in large Irish firms and the time spent developing a new product is linked with success in small but not large firms.

Large Irish Electronics Firms

Many of the characteristics of large Irish firms uncovered within the company level data are reinforced by the project level data. Large Irish firms develop mainly incremental new products that fit closely with their existing markets and product lines. Large firms are also seen to develop new products for markets rather than for specific customers. In common with small Irish firms, large Irish firms do not appear to be using project management techniques to develop new products, and, again confirming earlier findings, manufacturing issues are critical to large Irish firms

Small UK Electronics Firms

Small UK firms exhibit many of the characteristics of small Irish firms; they are still concerned with operational product development issues, for example product reliability; product development is a critical activity, and they target market gaps and niches. However, a few key differences between small UK and Irish firms were observed.

Small UK firms are closer to their customers than small Irish firms. This is probably due to the high level of export within the Irish firms. In addition, the small UK firms involved in this study did not appear to be as innovative in their NPD efforts as large UK firms – the opposite was found in the Irish firms.

Large UK Electronics Firms

As with the company level results there were few unexpected results relating to large UK firms. They were found to develop innovative new products and to be successful in high growth markets. Most of the findings relating to large UK firms were supported by previous literature.

1. The key findings on *R&D/Marketing Integration* are summarised below:

1.1 Larger, small firms (20 – 100 employees) appear to have difficulty in managing commercial and technical communications. This was not an issue for very small (fewer than 20 employees) firms or for large firms.

2. The key findings on *Project Management Skills* are summarised below:

2.1 Project Management Skills do not correlate with project success in Irish firms - results suggest that the project manager's role is not well defied in Irish firms.

2.2 Project planning and control is not linked with new product success in Irish firms.

3. The key findings on *Top Management Support* are summarised below:

3.1 Larger, small firms (20 – 100 employees) appear to have difficulty in managing top management support. This was not an issue for very small (fewer than 20 employees) firms or large firms.

4. The key findings on *Development Process Proficiency* are summarised below:

4.1 Small firms spend significantly longer at product development activities during unsuccessful projects than during successful projects. The length of time spent at development activities could be used as an early warning system for new product failure in small firms.

4.2 For all the firms in the study, more time is spent at predevelopment activities and less time at product introduction activities on successful NPD projects than on unsuccessful NPD projects.

4.3 The relative importance of the different NPD activities to small Irish firms suggests that they are involved in more innovative projects than large Irish firms.

5. The key findings on *New Product Characteristics* are summarised below:

5.1 Product reliability and compatibility are positively linked with success in small but not large firms. This reconfirms company level findings that small firms are still concerned with basic operational issues.

6. The key findings on *Market Characteristics* are summarised below:

6.1 Results relating to the impact of market growth and maturity on the success of products launched by small firms suggest that small firms avoid direct competition with large firms and that they seek market gaps and niches.

6.2 Success in small Irish firms is linked with developing innovative new products for specific/niche customer needs while success in small UK firms is linked with good customer relationships.

6.3 Two trends are indicated by the relationships between customer characteristics and new product success; (1) UK firms are closer to their customers than Irish firms, and (2) Large firms develop new products for markets rather than specific customers.

7. The key findings on *Marketing and Technological Skills* are summarised below:

7.1 Marketing skills are found to be more critical to new product success than technological skills.

7.2 Resource availability is a more critical issue for small than for large firms.

8. The key findings on *Marketing and Technological Synergy* are summarised below:

8.1 Large Irish firms develop new products that are close to their existing product lines and markets, i.e. incremental products

8.2 Small Irish and large UK firms are more successful with new products that are close to their existing product lines and markets – market synergy is a success factor.

9. The key findings on *Methods of Organising NPD* are summarised below:

9.1 Increased procedural formality is not significantly linked with success in any of the firms studied.

Table 8.21 Summary of Project Level Results.

8.4 Summary of INTERPROD Analysis

The INTERPROD data, at both company and project level, has provided considerable insight into the management of NPD in both small and large, Irish and UK firms. The inclusion of UK data up to this point in the study has helped to place the Irish results in an international context. The purpose of this section is to summarise the results that have emerged so far and to use these findings to update the research model.

Figure 8.1 shows the updated research model, indicating the factors that have been found to have an impact on the success of new products in Irish electronics firms. Because this model addresses only Irish firms, the impact of location on the development of new products is not shown in Figure 8.1

All of the variables tested were managed differently in small and in large firms; however, the management of five particular variables was not linked with new product success. It is particularly interesting to note that no link was found between *any* of the company-level factors and new product success. This was an unexpected result and will be explored further in the case studies presented and discussed in Chapters 9 and 10. Table 8.22 contains a summary of the relationships described in the model presented in Figure 8.1.

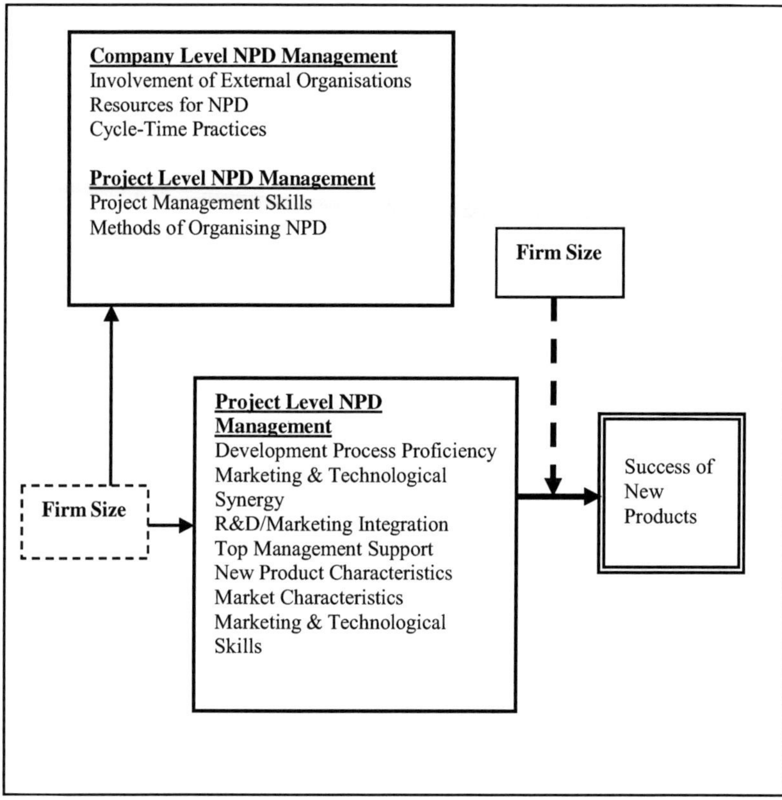

Figure 8.1 Factors influencing New Product Success in Irish Electronics Firms

Management Variable	Key Finding
Variables *not* linked to the success of New Products	
Involvement of External Organisations:	Small firms are more isolated than large firms in their NPD efforts although no link was established between collaborating with external organisations and new product success.
Resources for NPD:	Small firms allocate more resources to NPD than large firms, but there is no evidence that increased expenditure on NPD is linked with success in small or large firms.
Cycle-Time Practices:	Small firm use fewer project management techniques to reduce cycle-times, yet they are effective considering manufacturing in large firms is effective in reducing cycle-times. No link was found between NPD cycle-time and new product success, however.
Project Management Skills:	Unlike most other countries, project management skills are not linked with success in Irish firms.
Methods of Organising NPD:	No organisational method was linked with new product success – different organisational structures were useful in different situations.
Variables linked to the success of New Products	
Development Process Proficiency:	Proficiency in prototype development and market development and more time spent at predevelopment activities are linked with success in small firms. Manufacturing and marketing start-ups are linked with success in large firms.
Marketing and Technological Synergy:	Marketing synergy is linked with success in small firms but technical synergy is not linked with success in either small or large firms.

R&D/Marketing Integration:	Better integration increases new product success, particularly in firms with 20 – 100 employees.
Top Management Support:	Increased top management support increases new product success particularly in firms with 20 – 100 employees.
New Product Characteristics:	Product reliability and compatibility are linked with success in small firms.
Market Characteristics:	Innovative products for specific/niche customer needs are linked with success in small firms. Quickly growing markets are linked with success in large firms but not in small firms.
Marketing and Technological Skills:	Marketing skills and resources availability is linked with success in small firms.

Table 8.22 Key Findings Relating to Model Variables

The nature of this research is exploratory; the INTERPROD research tool has been used to examine the differences between NPD management in small and large firms. It has also allowed comparisons to be made between NPD management in Ireland and in other countries. The Irish INTERPROD data has been examined in two ways; in Chapter 6 the analysis presented in several INTERPROD publications was replicated in the Irish data; in Chapters 7 and 8 an in-depth analysis of the Irish and UK INTERPROD data was presented, concentrating on the differences between small and large firms. INTERPROD has been useful in exploring and describing the management of NPD in Irish firms, particularly small firms. The key points that have emerged from the analysis of the Irish INTERPROD data are summarised below.

CHAPTER NINE

CASE STUDIES

9.1 Introduction

In order to further explore the issues summarised at the end of Chapter 8 the management of NPD at three firms was investigated. These firms were:

- *eFlex Systems Ltd.*: A small firm with 9 employees who develops prototypes for third parties based on an advanced sensor technology that the firm has developed and patented.
- *Ferrostar Ltd.*: A small firm with 90 employees who develops and manufactures electronic instrumentation systems.
- *Telephase Inc.*: A large multinational manufacturer of precision high performance components with over 1000 employees at its Irish plant.

As discussed in Chapter 5, the three cases were selected with the following points in mind:

7.1 The Irish INTERPROD data contained one large firm for every 2 small firms; this ratio was applied in the selection of the three case-study firms.

7.2 The analysis of the INTERPOD data suggested that the management of NPD in very small firms (fewer than 20 employees) is different from that in larger small firms (between 20 and 100 employees). Therefore, one very small firm and one larger small firm were included in the case studies.

7.3 The three cases were selected to have different types of products and to operate in different markets. This will facilitate a deeper understanding of the impact of different variables on the NPD process.

These three firms have all completed the INTERPROD questionnaires. Section 9.2 provides a brief summary of the responses of these firms to INTERPOD. Sections 9.3, 9.4 and 9.5 present the case studies about each firm. The chapter is summarised in Section 9.6.

Each case study is laid out using the same structure. A brief background to the firm will be given followed by an overview of its NPD process. The management of NPD projects at each firm will be examined using five headings:

1. Cycle-time practices
2. Resources for NPD
3. Methods of Organising NPD
4. Involvement of External Organisations
5. Project Management Skills

These headings were chosen as they reflect some of the variables within the research model that have not been fully answered during the analysis of the INTERPROD data, see Figure 9.1.

However, the scope of the case studies is not confined solely to these five issues; the cases also help to explore the relationships summarised in the research model in Figure 8.1. A brief summary is also given at the end of each case. Throughout the presentation of the case studies, extensive use is made of direct quotations from those interviewed at the three firms.

The key findings emerging from these case studies are:

- The application and formality of NPD procedures varies significantly from firm to firm.
- The on-time delivery of new products is a serious issue for all three firms. Short or customer driven projects are easiest to deliver on-time.
- Even within the one firm the way in which new product projects are managed depends to a large extent on the engineers involved.

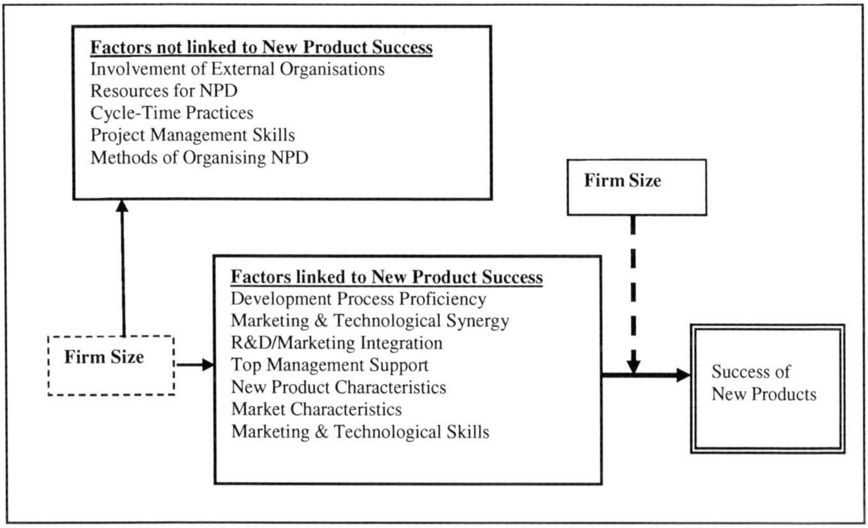

Figure 9.1 Research Model – Working Version

9.2 INTERPROD Results

Table 9.1 shows the overall performance data collected from the three firms involved in the case studies. It should be noted that three years elapsed between the collection of the INTERPROD data and the completion of the case studies. Therefore, the results presented in this section may not fully reflect the situation that prevailed in each of the firms when the case studies took place.

Some of the changes that took place in the three firms between the completion of the INTEPROD questionnaires and the case studies include:

eFlex: The firm changed management, reduced its staff to 9 and changed its strategy from developing and marketing products based on their technology to licensing and developing prototypes for third parties.

Ferrostar: Top management changed, the number of employees increased.

Telephase: The volume of product being developed and shipped increased dramatically.

Additional company level INTERPROD data for these three firms is shown in Tables 9.2 – 9.4. These tables do not contain all of the items presented in Chapter 7 but only those items that are considered to be relevant to the case studies described below.

Parameter	eFlex	Ferrostar	Telephase
Size [A6]	15	75	1300
Turnover (Millions) [A8]	£0.6M	£6.5M	£200M
Turnover per Employee (Millions) [A8/A6]	0.04	0.09	0.15
% Export [A12]	95%	100%	100%
% Turnover from new products [A21]	75%	30%	20%
% of employees in R&D (R&D Intensity) [A30/A6]	20%	27%	15%
Number of products available for sale.	5	4	5,000
Of the products considered for development over the last 5 years what percentage were launched? [L33]	20%	70%	70%
Of the products launched over the last five years what percentage succeeded? [L34]	50%	70%	60%
Average number of new products launched per year. [L36]	1.5	6	25
Our company is successful at *developing* new products. [L37]	Strongly Agree	Agree	Agree
Our company is good at *commercialising* new products. [L38]	Strongly Agree	Agree	Agree
Number of new products (not variants of existing products) launched during the last year. [IR20]	2	4	16
Number of new product variants launched during the last year. [IR21]	0	4	5
Number of patents held. [IR9]	4	3	100

Table 9.1 Overall Performance

Interaction with:	eFlex	Ferrostar	Telephase
Universities [A36]	Very Often	Never	Occasionally
Research Institutes [A37]	Occasionally	Never	Occasionally
Competitors [A38]	Never	Never	Never
Other Companies [A39]	Occasionally	Seldom	Occasionally
Customers [A40]	Very Often	Occasionally	Often
Suppliers [A41]	Seldom	Seldom	Often

Table 9.2 External Interaction

Spending on:	eFlex	Ferrostar	Telephase
Basic Research [A23]			5%
Applied Research [A24]	25%		
Process Improvement [A25]		15%	20%
New Product Development [A26]	50%	60%	70%
Product Quality Improvement [A27]		20%	
Cost Reduction [A28]		5%	5%
Marketing [A29]	25%		
R&D Expenditure as a % of sales [A33]	30%	13%	12%
R&D expenditure	£180K	£845K	£24,000K

Table 9.3 Product Innovation Spending

	eFlex	Ferrostar	Telephase
Average time from project initiation to commercialisation [A34]	6 months – 1 year	1 – 3 years	1 – 3 years
We rely on users to help us define and clarify their needs in NPD. [F6]	Agree	Disagree	Agree
How frequently do you contact customers about NPD? [F7]	Yearly	Yearly	Monthly
Our design changes grow more frequent as we move closer to manufacturing. [F22]	Disagree	Agree	Disagree
We often partner with our suppliers. [F23]	Strongly Disagree	Agree	Agree
Most effective cycle time reduction techniques:	Collaboration between R&D and manufacturing	Concurrent engineering and design	CAD/CAM and Simulation

Table 9.4 Cycle time Practices

	eFlex	**Ferrostar**	**Telephase**
Commercial Outcome	Met expectations	Slightly above expectations	Far above expectations
Technical Outcome	Met expectations	Slightly above expectations	Slightly above expectations
Technical Uncertainty	High	Low	High
Market Uncertainty	High	Low	Low
Level of information flow between commercial and technical entities	Low	Fair	Fair
Top management was supportive of this project	Strongly Agree	Strongly Agree	Strongly Agree
The market for this product was growing quickly	Disagree	Strongly Agree	Strongly Agree
The market for this product was mature	Disagree	Strongly Disagree	Neutral
Customers had a great need for this type of product	Strongly Agree	Strongly Agree	Strongly Agree
Customers' attitude towards new products	Resistant	Eager	Evaluative
We considered customers problems with previous products when designing this product	Agree	Agree	Agree
There was close fit between existing markets and the market for this product.	Agree	Agree	Agree
There was close fit between R&D skills and the needs of this product.	Neutral	Agree	Neutral
The project manager had the necessary management skills for this project	Agree	Strongly Agree	Neutral

Table 9.5 Project Level Factors (Successful Projects)

In completing the INTERPOD questionnaires eFlex and Telephase only reported on one successful new product; Ferrostar also reported on an unsuccessful new product. For consistency, Table 9.5 presents some of the parameters that relate to the management of successful new products in the three case study firms.

9.3 eFlex Systems

eFlex is a small electronics firm with 9 employees situated in the mid-west of Ireland. The company has developed and patented several advanced sensor technologies. The case study presented below was written based on interviews with five eFlex employees; these were the:

- Managing Director
- Engineering Manager
- Marketing Manager
- Development Engineer
- External Technical Consultant

The structure of eFlex is shown in Figure 9.2 below.

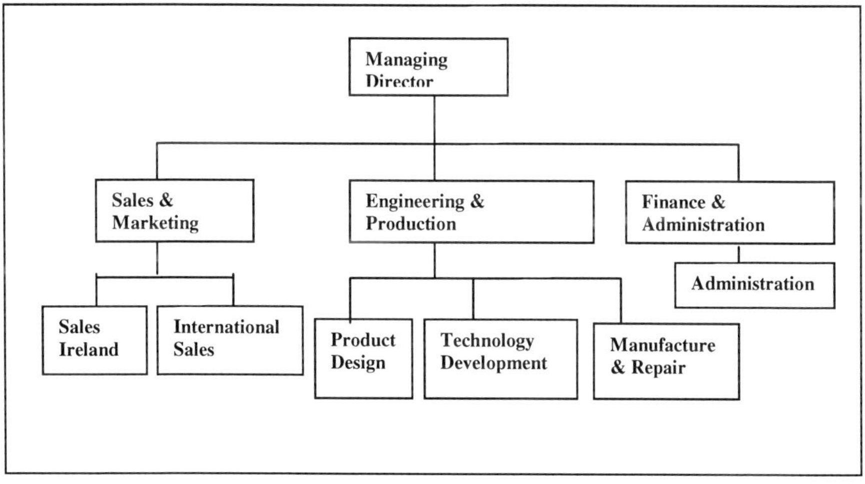

Figure 9.2 Company Structure - eFlex

9.3.1 Background

eFlex Systems has been operating for over 11 years. For several years after the firm was founded they developed, manufactured and marketed their own products based on advanced sensor technologies. During this period of time the firm employed up to 30 people. Development projects were large and costly. The focus was on the technology they were developing rather than on the market for their products. This period of time is described by one of the engineers as follows:

> We used to work under huge pressure 3 to 3.5 years ago, there was totally new technology, it was a nightmare. We couldn't meet schedules; there was animosity between sales/marketing and engineering. We didn't have the resources for the projects we were taking on.

Over 3 years ago the company experienced financial and managerial problems and underwent a fundamental restructuring. In an attempt to stabilise the company, eFlex shed over half of its employees and moved from developing, manufacturing, and selling its own products to its present strategy of developing prototypes and licensing technology to third parties. A new managing director was employed and he concentrated on working towards a break-even situation.

9.3.2 Overview of R&D Process

At the time that the interviews took place eFlex was in the process of moving from a detailed R&D process that was written to guide product development-from initial specification to product qualification- to a process that reflected the new operation of the company. eFlex is now in the business of providing a 'technical resource' to third parties. They use their advanced sensor technologies to develop prototypes for client firms.

Development projects usually start at eFlex as a response to a demand from a client company. A meeting is held between the relevant design engineer and sales and marketing from eFlex, and a representative from the client company. If the company is a new customer the MD of eFlex may attend. Additionally, if the project requires new or complex technology an external technical consultant who works closely with eFlex may join the meeting. The purpose of this first meeting is to get a clear definition of customer requirements. A lot of the initial contact about the design specification is done using e-mail, especially if the customer is located in another country.

NPD projects at eFlex are quite short; they concentrate on the front end of the development process. A typical project would be structured as follows:

a. Specification Phase.

b. Design Phase: simulate the circuit, layout the PCB and manufacture 1 or 2 boards.
c. Prototype Build: only one or two boards are built.
d. Testing: there is often an iteration between prototype and testing.
e. Shipment to customer for evaluation.

The NPD process at eFlex is not very formal but this is not viewed as a disadvantage:

> We do not have a well-defined process; there are no obstacles, other than people and on moving projects through quickly – one of the advantages of a small firm.

Projects are usually completed within 6 weeks, at which stage a working prototype is delivered to the customer. The other NPD phases including market research and business development, are undertaken by the customer.

9.3.3 Management of NPD Projects

Some the key features of how NPD projects are managed at eFlex are highlighted below.

Cycle-time practices

Projects at eFlex are normally 4 – 6 weeks in duration. At the start of a project the engineering manager sits down with the engineer involved and generates a schedule. This forms part of a project plan which is stored on the firm's network and is accessible to all employees. Engineers at eFlex do not use MS project or any other project planning software to generate schedules. The date for the final delivery of a product and project milestones are normally agreed with the customer. Project delivery dates are generally met – 'if they are missed it is only by a small amount'. When planning a project the engineering manager always schedules two design-build-test iterations; this helps to ensure that delivery dates are met.

Several characteristics of the NPD process at eFlex have contributed to their ability to meet delivery dates with new products. There is an 'increased focus on planning' and they also have 'three experienced people' developing products. They have a PCB layout facility in-house so that they can 'turn around PCBs in a few days'. eFlex have a bonus scheme for rewarding engineers for meeting schedules. Finally, in discussions about cycle times and meeting schedules, eFlex employees stressed the importance of agreeing schedules with customers, focusing on planning and meeting dates, and meeting regularly to review projects.

Resources for NPD

All NPD projects undertaken at eFlex are for specific customers. Most project costs are charged directly to customer accounts. While the interviews at eFlex were taking place a new system was put in place for tracking project costs; this appears to have worked well – '(we) have a system since before Christmas for logging cost and time...It's handy because at the end of a month it can easily give you costs to a particular account'. The largest cost associated with NPD projects at eFlex is engineering time; the company is 'driven by time so cost follows'. In summary, the management of NPD costs did not emerge as a critical issue during any of the interviews with eFlex.

Methods of Organising NPD

Weekly sales meeting are held at eFlex. The MD, the engineering manager and sales and marketing, all attend these meetings. These meetings are used to 'review issues from the market' and to 'focus on sales targets'. The meeting are described as being 'sales led but with an engineering presence'. In addition to these meetings both technical and marketing staff at eFlex describe technical-marketing communications as being very good. Two quotations help to illustrate this:

> There is a good relationship between sales and technical people – there is no clash, there is good understanding and cooperation.

> The commercial/technical interface is fine...No problems with the interface.

While there was little evidence of formal role definition, each group was quite clear about how they contribute to the NPD process.

> Engineering deals directly with customers, marketing may set up the first meeting but generally they are not the middle men – engineering takes control, the relationship is good.

> Marketing people may come up with new ideas, and ideas can also come from marketing within third parties; there is a good flow of information.

eFlex has only nine employees, the recently appointed MD is in the process of 'turning the company around'. He is involved on a weekly basis in NPD projects. He has a clear vision of what the firm needs to do in order to survive. At the moment eFlex are;

> ...focusing on 'fire-fighting' particularly on the sales side, trying to get to a breakeven situation – trying to stabilise the company

This vision is shared by others in the firm – 'there is now a focus on sales rather than technology, with an aim to return to a profitable position'.

However there is no evidence that top management involvement is turning towards interference, all those interviewed at eFlex appeared to be very supportive of the efforts of the new MD – 'its better at eFlex now, you can see where you are going', and '(the MD) is now turning around the company – it used to be very stuck'.

Involvement of External Organisations

One interesting feature of eFlex is their use of external consultants. Their core technologies have been developed with considerable assistance from an academic at a local university. This consultant is involved in almost all new products developed by the firm. He is an expert in the area of technology and has been working with the firm for many years.

At the start of a project the engineering manager sends the specification to the consultant who verifies the design using proprietary software that he has developed (in cooperation with eFlex). The consultant visits the firm at least once per week and also communicates regularly by phone.

eFlex also uses a sales consultant who has worked with the firm for many years and, like the technical consultant, is very knowledgeable about their area of business.

Project Management Skills

Despite the lack of a formal project management process projects at eFlex are tightly managed. A preliminary meeting is held at the start of each project, project schedules are agreed with the customer and these schedules are generally met. A system is in place to record all project costs and time, this is being used to control projects:

> Yes we do track project costs, that is why some projects are killed or put on hold. When working with a third party there is a budget drawn up and weekly or monthly expenses are reviewed, records are kept closely. We do track engineering time on projects.

9.3.4 Summary

eFlex is a small technology based firm that is currently struggling to survive. With only nine employees eFlex appears to operate as a cohesive unit. The current strategy is to license their technology and develop prototypes for third parties. This approach has meant that the NPD projects undertaken by eFlex are short and use known technologies. These projects are being delivered on time and staff at eFlex are satisfied that the firm is operating effectively.

9.4 Ferrostar Ltd

Ferrostar Ltd. employs 90 people and designs and manufactures electronics systems. The case study presented below was written based on interviews with nine Ferrostar employees, these were:

- Managing Director
- Development Engineering Manager
- Marketing Manager
- Production Manager
- Software Development Engineer (x2)
- Hardware Development Engineer (x2)
- Drafting Technician

9.4.1 Background

Ferrostar was founded by an Irish engineer over 20 years ago and, during the early 1990s, was acquired by a larger European company. The firm designs and manufactures electronic instrumentation systems. A substantial software package is required to control the systems developed by Ferrostar therefore the firm employs both hardware and software development engineers.

Ferrostar has two major product lines; 1) a customised system that uses both Ferrostar's own instrumentation and OEM equipment all controlled by proprietary software, and 2) a range of standalone instruments. The standalone instruments were not developed by engineers at Ferrostar's Irish facility but were introduced by the parent company during the early 1990s. In terms of the technology employed there is limited overlap between these two product lines. This means that the development department requires a broad mix of engineering skills – power, analog, digital, mechanical, software and firmware.

Ferrostar operates within a niche market in a mature industry. Products are sold both by an in-house sales force and through distributors. Systems sales normally require a significant investment by customers and therefore the sales cycle is long. This has the advantage that new product deadlines are not as crucial as they might be. However, late delivery of new products is still a problem for Ferrostar.

In the early days, before it was acquired by its European parent, Ferrostar operated like a 'typical' small firm. There were few procedures, projects could be rushed through, employees were highly committed and had complete control of how the firm operated. After the acquisition the operation of Ferrostar became more formal and many key decisions were made by the parent firm also the company grew larger. However, the culture at Ferrostar is still influenced by the early days as the quotation below shows.

> There is a lot of nostalgia about the 'old days' a lot of it relates to the small firm way of doing things, the ability to get things done. This is a double-edged sword; 1) it is possible to get things done quickly, 2) it creates a lack of respect for procedures and discipline. Sometimes the people who have a vision of how things can be done more quickly are actually blocking making things happen by not adhering to procedure. There are also some problems with the systems imposed by the parent company.

Ten years on the firm is still adjusting to its new status.

9.4.2 Overview of R&D Process

Since the acquisition in the early 1990s several different management structures have been applied to the product development department at Ferrostar. At the moment the firm has a well-documented NPD process, see Figure 9.3, but this is not considered suitable by senior development engineers – 'The procedures we use at the moment are too complex, they are too prescriptive, for example we have 10 pages where 3 or 4 would do'.

Despite this misgiving the process defined in Figure 9.3 is adhered to for most NPD projects. The management team hold quarterly Project Management Meetings, PGMT, where decisions are made about which new products start – 'There is centralised control over new products but its between an ill-defined group of people'.

The first step in the NPD process is the development of a Requirements List. This is supposed to be done by a Product Manager, however when the interviews took place at Ferrostar there was no product manager in place and the Requirements List was being drawn up by sales and customer support. Therefore the first stage of the NPD process is not closely followed:

> There is a deficiency in Product Management so that the first stages of NPD are sometimes blurred; adherence to procedure is about 50%.
>
> It's not working well, I don't think that anyone is happy with how it is working now – there should be a pull from the market but its not happening well now.

The Requirements List is reviewed at a Requirements Review – 'these work reasonably well, they fall down when people don't turn up and then complain when the final product doesn't meet what they expected'.

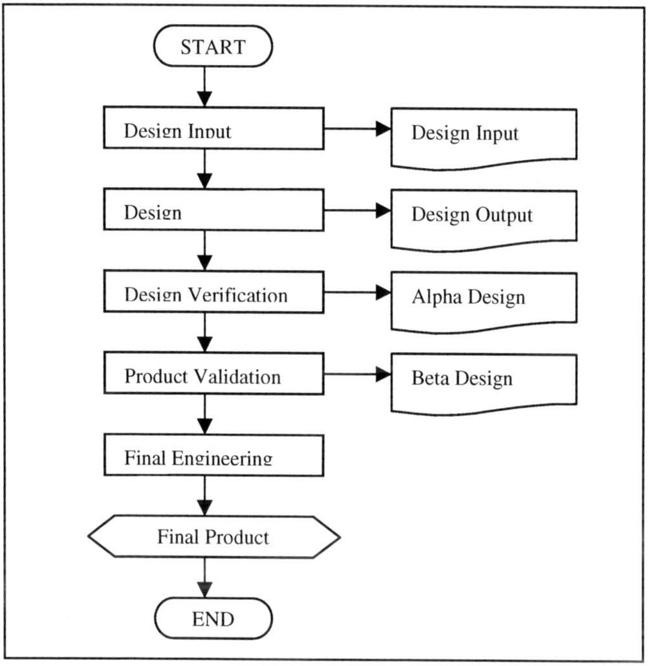

Figure 9.3 Ferrostar's New Product Development Process

The requirements list is the input for the development of a design specification this is the responsibility of the project leader who is appointed by the development manager. All engineers interviewed were happy with the manner in which design specifications were developed. However, the NPD process states that the project leader should work with a multi-functional group to write the design specification – in reality the specification is written by the project leader (a development engineer) alone. There is 'normally a formal meeting to get feedback on the design specification' but again this does not always happen:

> We try to have formal meetings to agree design specs.
>
> Design (engineers) write the design spec and run it by the product manager and marketing.

The next stage in the NPD process is the development of a project plan, again this is the responsibility of the development engineer. Project planning appears to be more rigorous for the hardware element of projects, where plans are developed and stored on the network. Software projects are only planned 'if there is hardware to go with it' and planning of software projects is less formal – 'we have white boards and draw schedules on them'.

The NPD process states that a Design Input Review should be held to review the requirements list, design specification and project plan. In general, it seems that the reviews outlined in Figure 9.3 do happen but it is unclear what actually happens at these reviews. For example, the following comments relate to a question about whether or not project plans are reviewed.

> No, we don't use review groups to review the project plans, if a plan is reviewed it would be with (the R&D manager) and the project leader.
>
> No – just design reviews....
>
> Yes, other engineers would look at the plan...- its informal.

Ferrostar's NPD process as shown in Figure 9.3 defines a number of formal reviews that should happen during the design implementation and verification phases however these reviews are not described in detail within the NPD documentation. The hardware development process at Ferrostar involves; component specification, circuit design and test, mechanical design, schematic capture, PCB layout, and prototype build and test. The software development process is not defined in as much detail within the NPD process.

The final release of a new product at Ferrostar usually requires the integration of hardware and software, complete testing (verification and validation) and the writing of sales, support and user documentation. This is probably the most difficult part of the process. System level problems often come to light during the integration of hardware and software. Also, the handover from R&D to Applications Engineering and production can be difficult – 'the handover can take almost as long as the project'.

9.4.3 Management of NPD Projects

Some of the key features of the management of NPD at Ferrostar are described below.

Cycle-time practices

Releasing new products on time is a significant problem for Ferrostar. A new product that is being developed for a specific customer it is more likely to meet its release schedule than a product that is internally driven:

> When it's for a customer order it's better but for a pure development project it's not particularly good. One or two high profile projects are years behind.

On-time delivery of new products was cited by many of those interviewed as a critical success factor:

> On-time delivery is the main (success factor) emphasised.

> Get a product that works accurately to customer requirements – on time

> ..timeliness is key, this is a major issue within the organisation.

> A successful product must hit production on time…

There was no general consensus about why new products were usually behind schedule. Some of the reasons suggested were:

> there is a reluctance to face up to how long things actually take, unexpected things turn up

> Also because other work has to be done – the R&D department provide technical support to other parts for the organisation

> On a recent project there was no stick or carrot for engineering to meet schedules – no negative implications for not meeting schedules. I think it could be 'carefully' incentivised.

> Implementation of JDE – it has added layers of bureaucracy to the process.

> Engineering resources in Ferrostar can be tied up fixing historical issues.

A question was put to those interviewed about the negative implications of missed deadlines. Some responses are listed below:

> If the release is missed it makes you look bad.

> The implication of being late is extra time and pressure to meet a shipment.
>
> Doesn't create a good impression – gives excuses to sales people – doesn't look good personally.

Yet despite an understanding that it is not good to miss release dates there is a culture of accepting that most projects are late. 'It is ridiculous how late some projects run e.g. one project should have taken 18 months but took 3 years.'

There is some evidence that there is an improvement in the timely delivery of software projects – this has been achieved by 'postponing features to future releases'. Unfortunately this strategy is more difficult in implement on hardware projects.

Resources for NPD

NPD costs are not controlled at Ferrostar. The total cost of the NPD effort at the company is budgeted and controlled on a yearly basis but there is no control over individual projects. Timesheets are not used to track project time and cost because 'no one looked at them when they were in use and the overhead was too high'. Additionally, the manufacturing cost of new products was highlighted as a problem – 'engineers have no concept of cost'. There appears to be no system for ensuring that the manufacturing cost of new products is controlled.

Methods of Organising NPD

Despite its small size there are communications problems at Ferrostar. The biggest problem is between engineering and production who blame 'each other for all problems e.g. building prototypes etc'. The engineering-marketing interface is better possibly because some of the senior marketing people have come from engineering. However, some stereotypes still apply; sales and marketing feel that R&D are too isolated and that they should get 'more clued into the customer and market place, they are not reactive enough'. While R&D feel that 'Sometimes sales and marketing are too focused on the current order and might push engineering to develop things that aren't cost effective'.

Information is freely available at Ferrostar – ‘the development department is quite open, and you can go directly to the engineer on a project and discuss a project’. But problems sometimes occur when people are not made aware of changes that are happening:

> In the past I have had major problems with spec changes not being communicated to sales and marketing.
>
> We get very little feedback from internal marketing, we get feedback by meeting customers.
>
> Decisions were made and agreed at a meeting but reversed at a subsequent meeting – there were no minutes circulated and not everything was minuted – I was at the meetings but my input was not always taken on board.

The regularity and formality of meetings varies from project to project but those interviewed were generally happy about how meetings were conducted. A few of those interviewed commented on the balance between technical and commercial groups in making decisions relating to NPD, the general consensus was that the situation had improved:

> Decision-making is fairly balanced between technical and commercial groups – previously it was more technically driven this was a problem with engineers changing specs.
> The requirements spec comes from the commercial people, there is a good relationship now, they have a good guy…
>
> Marketing probably have more influence, there used to be a problem, marketing was very far removed, didn’t make decisions, R&D had to decide what to do.

Top management meet once per quarter at a Product Management Meeting to decide what new product development projects should start. After this most of the decisions and the management input to the R&D process come from the R&D manager. The R&D manager agrees project schedules and specifications, selects project leaders and calls review meetings. Senior sales and marketing and customer support managers are also involved in the process providing inputs to product specifications and release dates.

Involvement of External Organisations

Ferrostar has made some use of external consultants in developing new products and their use is increasing. Typically they use consultants to take on part of a project or to 'answer fundamental questions'. External consultants account for approximately 10% of the RandD budget. Most of those interviewed felt that the use of consultants was an advantage to the company giving them 'access to specialist skills'. However, the disadvantage in using external consultants are that 'it is almost impossible to fully specify a new product so that you get exactly what you want' and 'it can end up costing more than required but this can also happen in house'.

Project Management Skills

The NPD process at Ferrostar describes the development of a full project plan including a task list, task dependencies, estimates of resources, time and cost of each task and a set of key milestones. This does not appear to take place. Project schedules are developed and stored on the engineering network – milestones are identified. Formal 'first meetings' are held for most projects. There are mixed opinions about the extent to which project schedules are maintained – engineering are happy with how schedules are updated, 'project plans are regularly updated, people will know if there is a delay' while others are not convinced, 'if the scheduling was adhered to we would know of the delays sooner, this would give the impression that there was some sort of control on the process'. There is also a problem with project specifications changing 'the sales people may change requirements half way through so specs need to change'.

Some of the engineers have tried using more formal project management techniques e.g. MS Project, but have not found them to be useful:

> We used to be big into MS Project but found it didn't work – we spent most time managing the schedule – managing projects at a minute level is a waste of time.
>
> I used (MP Project) in the past but it takes a lot of time…

9.4.4 Summary

Ferrostar has been through several changes in recent years, some of the problems experienced by the company today are a result of historical issues, but many also relate to resource limitations e.g. there is no Product Manager in place. A significant issue for Ferrostar is the on-time delivery of new products; some internal projects have been years late. The company now has a fully documented NPD process but not all projects follow this process. The management of projects varies from engineer to engineer.

9.5 Telephase Inc

Telephase Inc. is a leading manufacturer of precision high performance components. The firm has over 1,000 employees at its site in the mid-west of Ireland. The case study presented below was based on interviews with 13 Telephase employees, these were;

- Engineering Manager
- Product Line Director (x2)
- Senior Manager, Test & design Evaluation
- Product Line Manager (x2)
- Project Manager (x2)
- Senior Designer
- Test Leader
- Design Engineer
- Design Evaluation Engineer
- New Product Logistics Co-ordinator

9.5.1 Background

Telephase designs, manufactures, and markets a broad line of microelectronic components that address a wide range of customer applications. In 1996 Telephase restructured their product development activity along product lines. Prior to 1996 project teams were pulled together based on the project leader 'shouting loudest' for the resources required. Since 1996 each product line has become an individual unit. There are 13 different Product Business Units at Telephase in Ireland, each is almost autonomous; resources within each group are 'not available for sharing'.

Telephase is a multinational firm and each business unit can be physically located across several sites but a large proportion of design is located at its Irish operation. Two functions have been kept separate from the Product Line structure; these are Design Evaluation and Applications Engineering. Within these functions staff still report to product lines but they are located together. The primary reason for this is the cost of the equipment used by these functions, financially it would not make sense to replicate design evaluation and applications within in each of the 13 product lines. There is, however, anecdotal evidence that design evaluation works better than, for example, test engineering which has been divided by product lines.

At Telephase, in addition to the product line groups, different functional groups, referred to as faculties, have also been established. For example the test engineering faculty consists of 3 engineers who work across product lines.

At the moment there are between 120 – 150 new products under development at Telephase in Ireland. This could be reduced to 70 totally different parts as many products as very similar, differing only in the final phase of manufacturing. New products are classified as being:

1. Breakthrough – truly new
2. Platform – some changes but not completely new
3. Derivative – relatively minor change to an existing part.

The present portfolio consists of:

1. Breakthrough: 10 – 15%
2. Platform: 20 – 30%
3. Derivative: 50 – 60 %

9.5.2 Overview of R&D Process

Telephase's product development process is described in a document called CP-4.10. The process is divided into 5 different phases:

1. Feasibility
2. Implementation
3. Fabrication
4. Validation
5. Release

This is illustrated in Figure 9.4. As shown in Figure 9.4 there are seven milestones in the release of any new product at Telephase:

1. Design Start
2. 1st Lay-Out
3. 1st Boards
4. 1st Prototypes
5. Final Layout-Out
6. Final Boards
7. Release

In addition, five CP Review Documents (CPRs) are identified. The first one of these, CPR 0 Review is only necessary for breakthrough of platform products. A CPR is a business review of the new product to be developed. CPR 1 is considered to be the start of a project, in other words, projects should not start until CPR1 has been signed off. In reality some projects do proceed without a CPR 1. CPR 2 occurs after 1st tape out, that is when the product has been designed but not yet manufactured into an actual part. CPR 3 is the final sign off before the release of the product and CPR 4 happens typically 6 quarters (18 months) after the product has been released and examines whether or not the product has been a success in the market place.

Telephase has a full time NPD process manager who is responsible for their NPD process – he has three co-ordinators working with him. They are responsible for ensuring that CP-4.10, and other related procedures, are kept up to date, that any changes are communicated to all relevant personnel and that training is provided where necessary.

The process outlined in Figure 9.4, CP-4.10, is followed for every NPD project at Telephase. But not all development engineers are entirely happy with the process:

> Everyone knows CP-4.10, we have to follow it, we are doing it because we have to, not because it is the correct thing to do.
>
> I think that there is probably too much adherence to CP-4.10, maybe there should be a different process for Breakthrough, Platform, and Derivative projects. For example the amount of testing that is done on derivative parts, maybe yield analysis should only be done on the new circuitry.
>
> We follow CP-4.10 out of fear of audit

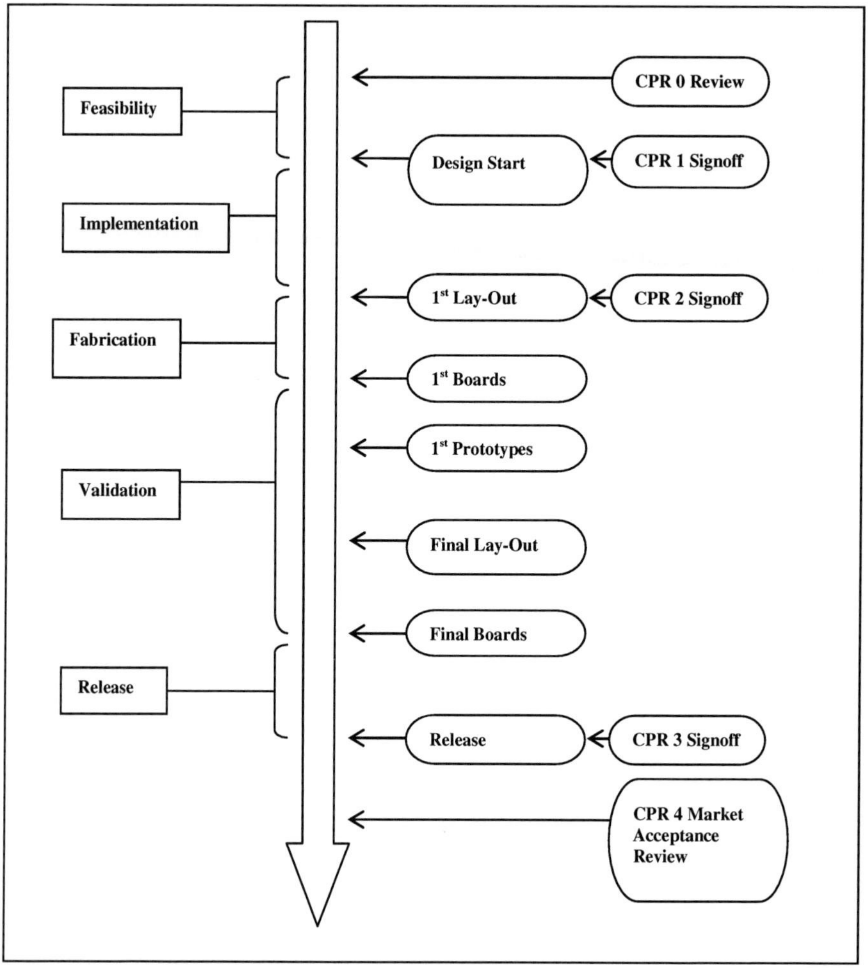

Figure 9.4 Telephase's New Product Development Process

One of the problems with the process is the rate at which CP-4.10 is updated, 'it is now at revision AX', it can be difficult for engineers to keep up with the changes. Another problem, also identified above is the difficulty in having one process to cover a broad range of NPD projects, from minor changes to breakthrough products.

A key issue for the management of new products at Telephase is the predictability of the process. There is a feeling that the process is not in control. Engineers complain of being over worked and not being able to control their workload – 'a very high number of resources are over committed'. One senior manager summarised the problem by saying that 'Telephase does not know the capacity of its NPD engine', in other words they do not know how many projects can be handled effectively at one time.

9.5.3 Management of NPD Projects

Some of the key features of how NPD projects are managed at Telephase in Ireland are summarised below.

Cycle-time practices

Some phases within the product development process at Telephase are very predictable but others can vary significantly. A typical time line for developing a new product is shown in Figure 9.5 below.

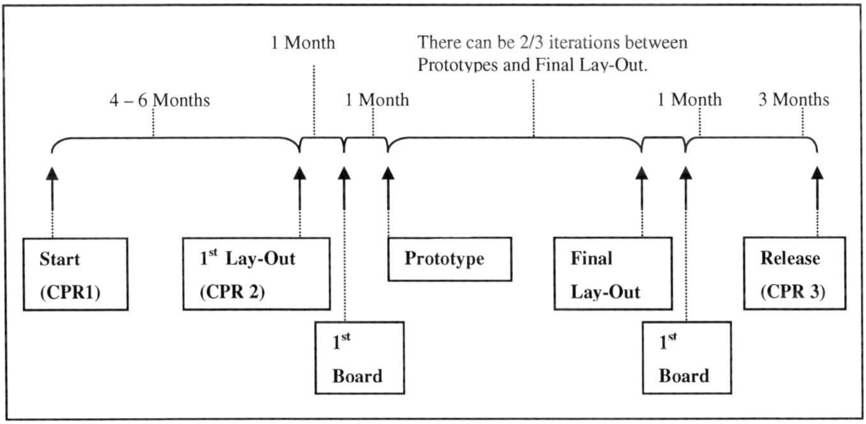

Figure 9.5 Typical NPD Project Time Line

Most new products at Telephase take 12 – 18 months to release, but there is little confidence in Telephase about the ability to meet NPD schedules:

> We do use MS Project – but generally milestones are not met and people are not worried about it, it is only a problem if the customer has been given a date.

> Milestones by definition are not adhered to!

> Yes, there are milestone charts for each project but they are not normally adhered to.

> Yes, we develop an MS Project schedule for each project. We don't often meet the schedules – they tend to be a bit ambitious at the start and schedules slip a lot. We plan on the best-case scenario. If schedules slip it doesn't really seem to bother top-level management as significantly as it should.

The main problem is the length of time between the release of first samples and the final tape out, see Figure 9.5. The 'tape out' milestones refer to the completion of the design, using CAD, by the design engineer. From the time when samples of a part first become available to when the design is complete there can be several iterations, these are the main cause of delays within the NPD process at Telephase. Reducing the number of iterations on design projects is a major concern for Telephase's senior engineers. They have established a RFI (Reason For Iterations) project team that has come up with the following recommendations:

- More detailed simulations of designs before going to tape out.
- More detailed evaluation of parts before tape out.
- Continued role of the RFI team.
- Establishment of 2 new advisors to assist in the simulation and evaluation processes.

Several of the engineers interviewed also felt that frequent iterations on projects were often caused by carelessness:

> These iterations could be reduced by doing things right day one…

> Far too many iterations – Not enough time at the Design stage to simulate – some of the mistakes are ridiculous

Another issue that emerged is the lack of comprehensive, realistic planning of NPD projects:

> Time planning – not in enough detail, should use the team but not forming teams early enough and not getting buy into the schedule.

> No one accepts the time it will take. Difficult to get support from others e.g. Probe. Often wonder if I would be better off doing 2 hours 'work' rather than prepare the schedule.
>
> We are good at meeting schedules when there is one large lead customer driving the schedule, but poor with multi-market parts
>
> Are schedules taken seriously? If groups are meeting everything then the schedules are probably padded.

The final comment above from one of the senior managers interviewed highlights another issue. Design engineers feel that management impose unrealistic schedules – 'Design Schedules usually are unrealistic, try to make up for it at the back end, make mistakes rushing and end up slipping further'. But when engineers try to put in place what they believe to be more reasonable delivery dates these are rejected by top management as being 'padded'.

Resources for NPD

Engineers at Telephase are not concerned about the cost of NPD projects – 'Project cost is not something that most of the individual engineers on the team are very involved with'. It is not easy for managers to monitor project costs, a comment from one manager who tried to closely monitor the costs within his group illustrates this:

> I think that we need a better project cost system, I am now capturing 60% of total project spend, i.e. I don't know where 40% of my costs are going. I think that we are losing about 30% of people's time – where is it going?

Even at a more senior level within the firm the time it takes to develop a new product rather than the cost is the main issue:

> Telephase is so dominated by time and iterations that this is what is really meant by cost.

Methods of Organising NPD

The regularity and quality of meetings at Telephase depends on the manager or team leader, though in general meetings are held weekly and minutes are issued electronically to all concerned. Some of the comments below reflect this situation:

> Weekly new product meetings. Always start with the design people who go off on tangents on technical discussions that should take place off-line

> Weekly, agenda, minutes, focused, short. The project manager interfaces with Marketing who sits beside him; more involved at feasibility, bolted into the team; circulate minutes.

> Yes, we have weekly meeting, they tend to focus almost exclusively on technical issues rather than on project management – it is very much personality driven..

> The group does hold weekly meetings but these can be cancelled depending on the phase of the project.

The restructuring of the R&D department into Product Lines had the effect of limiting the flow of knowledge between engineers. Information is more likely to flow within a product line group that between them:

> Since we went to Product Lines – we have lost faculty knowledge – if it's not within your own group forget about it – even if the expertise is sitting beside you. Could have a guy in from Teradyne that doesn't get shared – a sense of competition developing that's not healthy.

> Within the group there are experts, they are not part of the design team, help is also available outside the group it is all pretty informal.

For example, design reviews are supposed to be attended by engineers from different product lines in order to facilitate learning particularly by younger engineers. However unless the lead engineer puts effort into ensuring that there will be a good attendance, numbers can be low.

> Design review for all design faculty but don't happen early enough and not enough people (from outside) come to these.

> The Design Review is the key Risk Management activity. Not enough people attend unless the designer goes and asks people. Some people give good reviews and people will attend because they learn from the experience.

Finally, the commercial/technical interface does not appear to be problematic for Telephase, as the comments below illustrate:

> No problem with the technical/marketing interface. Marketing are not at team meetings, they are called in when needed.

> Marketing people attend meetings ... In general the relationship between marketing and engineering is quite balanced.
>
> The commercial-technical interface is not terribly problematic – I have learnt to avoid creeping elegance, I 'push back' more e.g. if marketing ask for an additional feature I may say 'yes but it will add time to the schedule'. Marketing do attend weekly meetings but sometimes the engineers are selective about what they are told.
>
> In some cases the technical/commercial interface is very well managed.

Every Monday morning senior management at Telephase meet to review NPD projects. Projects are presented to a core group of product line directors and senior sales and marketing executives. CPR1 Business Reviews are conducted at these meetings and the senior management group can ask the development team to change what they propose to do. However, they rarely 'kill' projects.

> Reviews are held every Monday morning, religiously, we carry out many reviews, there is a core group on the CPR review board. Projects are seriously questioned, engineers are often told to go away and review/revise their plans – this is a substitute for killing projects.

Telephase also run annual planning meetings that are used to define the NPD targets for the coming year. Beyond these two meetings the involvement of senior management in NPD is most likely to occur when a key customer is looking for a new product.

The management level below Product Line Directors is made up of Product Line Managers. This level of management at Telephase is involved in NPD on a daily basis. The NPD procedures at Telephase define the role of a Project Sponsor; in general this role is fulfilled by the product line managers

Involvement of External Organisations

Telephase sometimes interact with local universities 'for masters projects to try out new ideas' but they don't normally use external consultants:

> There are no external people involved in developing products.
>
> There isn't much use of external contractors.

Project Management Skills

Since Telephase reorganised into Product Lines five years ago different management styles have emerged in different groups. All NPD projects are planned and controlled but the manner in which they are managed varies from group to group:

> Use 7 milestones from the CP document. XXX have a top level MS Project chart for all projects … The group do try to keep to milestone dates – pretty good at meeting schedules, of the 8 active projects 5 are on time, meeting dates.

> A schedule is put in place at the start of each product, covers the key milestones and time allowed. Depends on the team leader. The first schedule is realistic – then asked by product line manager to add in something here and there, and suddenly you are far too tight.

Some of the groups at Telephase are using dedicated project managers and are reporting increased success in delivering NPD projects within schedule. One director interviewed agrees that:

> Dedicated project managers indicate the way for the future but in the past Telephase has been unsuccessful at imposing disciplines on the engineers.

The manager of the NPD process also agrees that the firm needs some level of project management – 'How can you manage €40M of R&D without Project Management?' However the problem is in implementing new work practices. There is a history within the firm of new management techniques being introduced, training being provided and engineers then returning to do exactly as they did before. Telephase is 'a democratic engineering company' where it is very difficult to impose new practices on the engineering community. Additionally, technical skills are very highly rated and project management is not seen to be a 'technical' issue. Designers consider managing projects to be largely an administrative issue and are not interested in becoming involved - there is 'no strong project management culture' within the plant in Ireland.

However, there are two full-time project managers working in R&D at Telephase, both are design engineers who, with the support of their product line managers, have taken on the role of project management. The firm is struggling with how to integrate this new role into their existing structure:

> Career path is a big issue in introducing project managers, also the difference between project managers and project leaders, i.e. the technical skills vs. the management skills.

Initial evaluation of the projects that have been managed by these 'full-time' project managers shows that they are being completed in less time than similar projects that didn't use project managers. Also, the project managers interviewed reported higher success and more satisfaction with meeting deadlines than other engineers:

> I am using teams to agree the schedule, mainly with the designer and a bit of input from the test person. Since last November I am setting realistic release dates.
>
> Walk through current activities on the plan at the weekly meeting – even if it only takes 5 minutes. Don't want to go back to where we only had meetings if there was a problem.
>
> Marketing attend all weekly meetings. Get business updates from marketing regularly – great motivator for the team.
>
> The group places huge value on the Project Plans – completed based on the teams' inputs, lots of discussion.

9.5.4 Summary

Telephase is a large firm with a detailed NPD procedure and more than 120 new products under development at any point in time. The firm is organised around product lines and each product line operates with a high degree of autonomy. Management at Telephase are concerned that their product development 'engine' is not under control; most new products are not released on time. Full-time project managers are working in some of the product lines and initial evaluation shows that they have increased success in meeting NPD schedules.

9.6 Conclusion

The management of NPD at three of the firms who participated in the INTERPROD study was examined in more detail in order gain a better understanding of some of the issues that emerged from INTERPROD. The firms were selected based on their size and the nature of their operations.

NPD projects are managed quite differently at each of these three firms. In particular differences are seen in how the firms:

1. Apply formal NPD procedures
2. Manage inter functional communications
3. Plan and control NPD projects.

CHAPTER TEN

CASE STUDY DISCUSSION AND KEY FINDINGS

10.1 Introduction

This chapter contains a discussion of the three case studies presented in Chapter 9 and also the key findings of both the survey and case studies. Section 10.2 will contain an analysis of the three cases highlighting the differences between them, it will be organised to reflect the structure used in presenting the cases. The case study findings will be summarised in Section 10.3. The research model will be updated for the final time and used to propose a series of hypotheses about the management of NPD in small firms in Section 10.4. Key findings from both the INTERPROD survey and the case studies will be presented in Section 10.5. Finally, the conclusion to the chapter will be in Section 10.6.

10.2 Explanation of Research Variables

The purpose of this analysis is to seek explanations for some of the relationships identified in Chapter 8 and also to describe in more detail the management of NPD in small Irish electronics firms. Each of the three firms described in Chapter 9 is operating in a different situation, using different technologies for different types of products offered to different customers. Previous studies (Loch, 2000; Bessant & Francis, 1997) would suggest that the NPD processes implemented in each firm should also be different. The discussion below will compare the NPD process in place at each firm and use the case study data to offer an explanation for some of the relationships proposed by the research model.

The NPD processes employed by each of the three firms vary significantly this reflects the various contexts within which these firms develop new products. The main characteristics of each firm are described below.

- eFlex is the only one of the three firms that is not involved in manufacturing. The firm operates as a design house developing customised prototypes based on a core technology. eFlex appears to have a tightly managed process but they develop only incremental new products. eFlex also has a limited product portfolio; each NPD project at eFlex is an extension or application of their one core technology. When eFlex were developing finished products they were not managing well, in fact the firm could not have survived following this strategy and underwent fundamental restructuring.

- At the other extreme, Telephase is a large multi-national with over 5,000 products in their portfolio. The fact that Telephase is a component manufacturer accounts in some way for the large product offering of the firm. However even taking this into consideration Telephase, as was shown in Table 9.1, is significantly larger than either of the other firms. Telephase have a complex procedure for developing new products that facilitates some degree of flexibility. Telephase have a team whose sole function is to manage the NPD process, for cost reasons it would not be possible to implement this at either Ferrostar or eFlex. The NPD procedure at Telephase is adhered to but this is partially because it is policed – the NPD process manager ensures that all new products are developed in accordance with CP-4.10.
- Ferrostar is the only firm that develops a significant amount of software to operate the hardware products they supply. This means that Ferrostar has a broader mix of engineering skills than either of the other firms. Ferrostar appears to be having most difficulty with product development. The firm has a detailed NDP procedure but the application of this procedure is not consistent, reviews are held on some projects and by some developers but not by others. At Ferrostar it is not clear who is responsible for ensuring that the NPD procedure is followed.

The cases show that neither of the small firms, eFlex or Ferrostar, have implemented and adhere to a detailed comprehensive NPD process. The process at the large firm, Telephase, is not perfect but there is a much stronger impression that it is under control.

The differences in the R&D processes used by each of the three firms studied shows that, as suggested by several authors, there is no one 'best' way to manage NPD. The complex procedure in place at Telephase would not work in the smaller firms and the team approach seen at eFlex would not work in either Ferrostar or Telephase. An examination of the differences between the firms in terms of how they manage the five variables examined in the case studies follows.

10.2.1 Cycle-Time Practices

Completing NPD projects on time is not a problem for eFlex, but they are developing applications or incremental products rather than totally new products – previous research has shown this to be an effective strategy for reducing time-to-market (Ali et al., 1995). Additionally, the survey results presented in Chapter 8 identified a link between incremental products and shorter NPD cycle-times.

Telephase have a problem with meeting product development schedules however they are taking definite action to solve this problem. For example, the RFI teams show that they are adopting a *compression* strategy (Eisenhardt & Tabrizi, 1995) rather than an *experiential* strategy to reduce development time. This is probably appropriate given the incremental nature of a lot of their products but may not be suitable for breakthrough products. Telephase are also examining the use of more formal project management techniques to improve their time to market. They have implemented changes (full time project managers) and are measuring the impact of these changes. Focusing on time as a goal, as they are doing, is likely to produce positive results (Zirger & Hartley, 1996).

Ferrostar also have problems with NPD time management but do not appear to have any definite strategy for solving these problem. Because of the type of product that they are developing, capital equipment with a long sales cycle, their customers appear to accept longer delivery schedules. However, longer NPD cycle times will cost the firm in terms of both resources and missed opportunities.

These findings help to explain the results from the INTERPROD survey. The lack of a link between cycle-time could be explained by the fact that most firms are trying to move faster but different sectors of the electronics industry move at different rates. Firms in the faster sectors simply have to keep up or they cease to trade, for example the change in strategy at eFlex or the positive actions taken by Telephase. In the slower sectors time-to-market is not a critical issue and so is not strongly linked with success.

10.2.2 Resources for NPD

Only one firm studied, eFlex, was closely monitoring NPD costs. This is linked to the way that the firm operates. All the new products that they develop are paid for by customers, eFlex needs to be in a position to accurately calculate NPD costs.

At Ferrostar there appears to be little control over project costs or the time being spent on particular projects. A time keeping system that had been in place was discontinued because it was not being used and there was no evidence of a new cost or time control system being implemented. Since there is so little awareness of the cost of developing new products it is likely that Ferrostar are overspending on their development activities. This could well be the case in other small firms.

Costs are not closely tracked at Telephase but there is a much greater awareness of the amount of time spent at different projects than there is at Ferrostar. At Telephase 'time is money' and time is closely monitored. The NPD process manager keeps a record of how long projects actually take to reach each milestone compared to the scheduled time. This information is used by other managers to set the overall product development schedule on an annual basis. The information is also used to evaluate the performance of the different product lines. Telephase is not typical of the large Irish firms included in the INTERPROD survey. Their expenditure in R&D, see Table 9.3, is significantly higher than the mean for large Irish firms studied, see Table 8.10. Therefore the Telephase case study is not useful in explaining the findings of the INTERPROD survey that large Irish electronics firms do not invest heavily in NPD.

The lack of control of NPD costs at Ferrostar could help to explain the lack of a link between increased NPD expenditure and increased new product success in small Irish firms reported by the INTERPROD survey. Small firms need to control their NPD expenditure before they can expect to gain a benefit from increased expenditure. It is likely that firms who spend more are less efficient at NPD. This also questions the use of R&D expenditure as a measure of effective NPD activity especially amongst small firms.

10.2.3 Methods of Organising NPD

With only 9 employees eFlex is small enough to operate as a cohesive team and this appears to be what is happening. There are few formal structures, clear role definitions and few problems with communications. NPD at eFlex is under top management control. The managing director has a vision of where he wants the firm to go and this is supported by the other staff.

Ferrostar has problems with communications; relationships between different functional groups are difficult. Also, the firm has changed its managing director several times in the last few years. NPD at Ferrostar is organised under an R&D manager who is responsible for hardware and software development teams. Despite the difficulties between the different functional groups individuals still share information, though not formally.

At Telephase NPD is organised into product lines. Each product line has a manager who has day-to-day control of NPD. The way in which NPD is managed varies from product line to product line, and different practices often apply. However, communication and the sharing of information relating to NPD is largely defined within the NPD procedure. Also, no serious problems with the R&D/marketing interface were reported at Telephase.

Examining these three firms would appear to suggest that the method of organising NPD depends on the context of the firm and the nature of the product. The 'best' structure to apply is the one that suits best. It is therefore not surprising that no particular structure was linked with success in the Irish data. The cases also support the survey data relating to NPD process formality. As a matter of necessity, because of the increased complexity associated with a much larger product portfolio and also because of the number of R&D personnel, Telephase have a detailed and formal NPD procedure. The smaller firms can operate without such formality.

10.2.4 Involvement of External Organisations

Only one of the firms studied had regular contact with external organisations, this was eFlex. In this situation there was a very close relationship between the firm and a university-based consultant. This relationship was extremely beneficial to the firm.

Ferrostar reported some involvement of external consultants in their NPD projects but their experiences reflect some of the published literature (LaBahn et al. 1996; Rothwell & Dodgson, 1991). They found that it was difficult to manage relationships with external consultants; costs were difficult to control and consultants did not always deliver what was required. Telephase reported very limited usage of external resources.

The case studies support the survey findings. There is not extensive use of external resources for NPD; only one of the three firms regularly used outside help in developing new products. Additionally, the experiences of Ferrostar describe some of the difficulties that firms encounter in interacting with external organisations on NPD projects. Within the three case studies the degree to which the use of external resources is linked with success depends very much on the particular situation. This is a complex issue, the survey and cases provide some insights into the involvement of external organisations in NPD projects but a more detailed study would be required to fully address the topic.

10.2.5 Project Management Skills

As with other management practices the relatively uncomplicated situation at eFlex means that its projects are well managed. At Ferrostar there is evidence that projects are being planned but there is no consistent project management process. Some of the engineers at Ferrostar have attempted to use more formal project management techniques but have not been successful.

Management at Telephase are exploring the use of project management. The firm already schedules projects in some detail, producing project plans and using these plans to monitor progress. The company is now exploring the possibility of putting in place full time project managers, many of the problems in doing this relate to the 'engineering' culture at Telephase.

None of the firms studied used detailed project management structures, none of them had clearly defined the role of the project manager within their NPD structures. This supports the finding of the survey that project management techniques are not widely used in Irish firms. The survey also found that increased project planning and control were not linked with success, the Ferrostar case provides some explanation for this, detailed planning of projects is not always beneficial and can delay a project by wasting resources.

This appears to be an issue that is peculiar to Irish firms. The INTEPROD survey showed that project management was linked with success in UK firms and also engineers from Telephase reported that project management was more commonly used in their UK facility.

10.3 Summary of Case Studies

The three case studies presented in Chapter 9 have explored in further detail some of the relationships and trends identified by the INTERPROD survey. The INTERPROD survey provides a broad picture of how NPD is managed in small Irish electronics firms and also of the success factors that apply. It helps to identify broad trends and compare the management of successful and unsuccessful new products. It is limited however in its ability to explain the relationships and success factors identified. The case studies on the other hand have been useful in their ability to explain the relationships and factors identified by INTERPROD but they are cannot be used to generalise to the total population. Therefore, in the discussion that follows conclusions can only be drawn about the three firms involved in the case studies.

The main findings from the case studies are summarised below.

1. *The large firm in the study was in a position to devote considerably more resources to the development and maintenance of a NPD process than either of the small firms.* The large firm had a team of people dedicated to managing their NPD process. This meant that there was sufficient 'policing' of the process to ensure that all projects followed it. It also meant that when problems with the process were encountered, for example the length of time taken to develop a new product, they could assign people to address the problem. When the large firm decided to investigate a change to their existing process, i.e. the use of dedicated project managers, they were able to try out the new method and evaluate its impact on NPD performance. Neither of the small firms in the study were in a position to dedicate staff full time to maintaining their NPD processes and so it was more difficult to ensure that they were adhered to or that they were updated regularly to reflect the changing needs of the firm.
2. *Only the smallest firm kept regular and accurate control of their NPD costs.* Because they were operating as a design house and all their product development costs were billed directly to their customers the smallest firm studied had to closely track their NPD costs. Neither of the other two firms accurately controlled

how much they were spending on NPD. In the case of the large firm, time was closely monitored and this could be translated into cost. However, in the case of the second small firm there was little control over what it cost to develop a new product.

3. *Each of the three firms took a different approach to organising for NPD and each firm appeared to have an organisational structure suited to their circumstances.* The way in which NPD was organised varied greatly from firm to firm. The largest firm was organised into product lines, the next firm into an R&D department and the smallest firm had little formal structure but operated as a venture team. In each case the structure used was suited to the size and operation of the firm.
4. *The importance of time-to-market depended on the market in which the firm was operating.* The smallest firm studied was involved in small, fast customer specific projects which were normally delivered on time – time to market was not a problem. However, neither of the other two firms was successful at releasing new products on time. In the case of the large firm that was operating in a competitive, rapidly changing industry sector, this was a significant problem and was being addressed. But the second small firm had not made any serious effort to reduce the length of time it took to release new products, this could be a reflection that the industry in which they were operating was not driven by time.
5. *The degree to which external organisations were involved in NPD and the benefits gained from this involvement varied from firm to firm.* Only the smallest of the firms studied had any significant involvement with an external organisation. This was a close relationship that had been developed over a number of years. The second small firm had involved external organisations but was not happy that the collaboration had been successful. The large firm did not involve external organisations in NPD. There was no clear link between the involvement of external organisations and success or firm size.
6. *The nature of the NPD process and the type and quantity of new products being developed varied with firm size.* The NPD processes that were in place varied from firm to firm, as did the type of product being developed. The smallest firm, eFlex, had attempted and failed to develop a complex product from ideation through to commercialisation. It was only when they limited the range of their NPD efforts that they began to experience some

degree of success. The middle-sized firm, Ferrostar, were developing a relatively complex range of products requiring a broad range of skills but they were struggling to succeed. There is a very clear impression from the case study that they were overstretched. It is interesting that at an earlier time when Ferrostar had a more limited range of products they experienced higher success levels. The large firm, Telephase, developed a wide range of products at the forefront of technology and had a very large team of technical experts.

There is a sense then that the extent of product innovation that can be undertaken by a firm varies with its size. Here again it is possible to speculate that the 'adolescent' or mid-sized firm has the greatest difficulty when they try to extend their product range and technology base they face high risks. For example, eFlex tried to do this and eventually retreated back to a smaller size with a less complex NPD requirement. Again this suggests that it is as firms grow that they experience most difficulty in managing NPD.

While the summary of the case studies cannot be generalised to the population of small electronics firms in Ireland, case studies are useful in developing theory. The theory that has been developed from this research about how NPD is managed in small Irish electronics firms is embodied in the final version of the research model presented in Section 10.4, and the accompanying hypotheses.

10.4 Key Findings from INTERPROD and Case Studies

The case studies have provided a deeper understanding of the relationships shown in the research model and allow explanations to be offered for some of these relationships. As stated in the introduction and reiterated throughout this thesis, this project is exploratory, the objective to build theory about the management of NPD in small firms. Therefore, a suitable way to summarise the results is through the proposition of a set of hypotheses that describe the relationships presented in the final version of the research model.

While the analysis of the case studies has provided some explanations for the relationships proposed in the research model it has not resulted in any change. At the end of Chapter 8 a model was presented showing seven variables that were linked to new product success and five that were not.

All variables were affected by firm size. No evidence has emerged to question the relationships between the seven variable identified and new product success. Also, the case studies have provided some explanations for the lack of any relationship between the five variables and new product success. The final model is shown in Figure 10.1. The twelve hypotheses are discussed below.

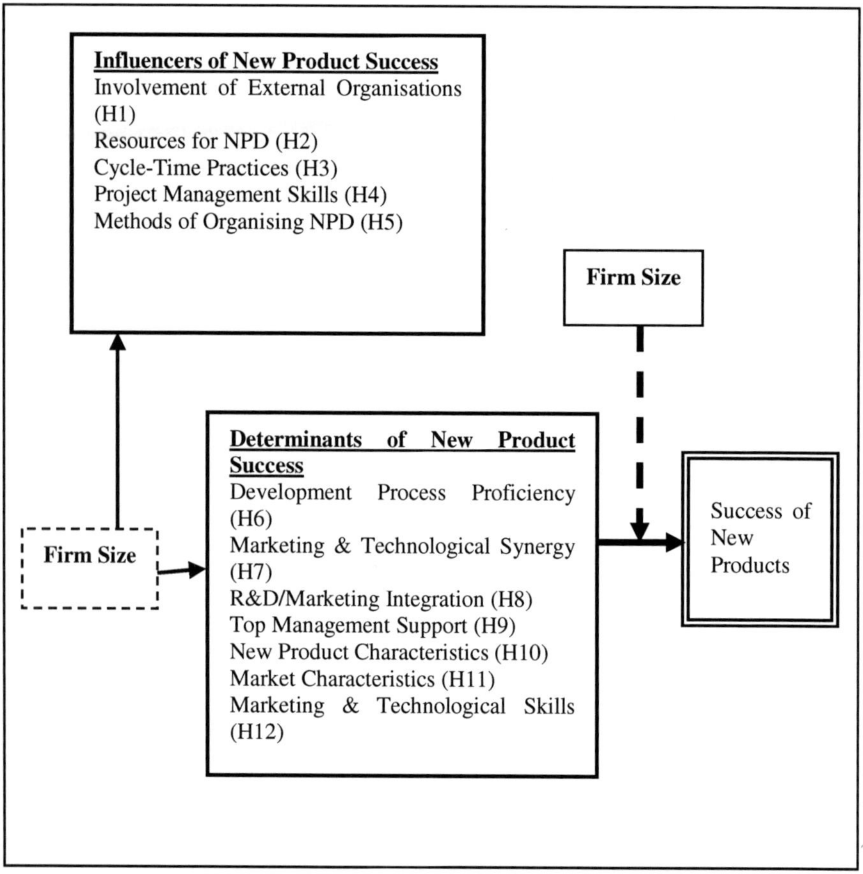

Figure 10.1 Research Model – NPD Management at Small Irish Electronics Firms

H1: The involvement of external organisations in NPD projects in electronics firms is not linked with increased new product success.
Rationale: The electronics sector is based on a complex and highly differentiated technology. It is difficult for firms to find external consultants or organisations that have the technical skills and knowledge to contribute to their NPD projects. Additionally, even if the required skills are available external to the organisation the process of managing the relationship with an external entity can be difficult for small firms in terms of controlling deliverables and costs. Support for this hypothesis comes from the INTERPROD survey that found that the involvement of external organisations in NPD did not have a significant positive impact on project success and could be detrimental to new product success. Additional, support for the hypothesis comes from the case studies; of the three firms studied only eFlex regularly involved external organisations in NPD projects. Ferrostar also used external resources for NPD but had difficulties with the relationship. There is also support for this hypothesis in the literature surveyed (Love & Roper, 2001; LaBahn, 1996).

H2: Increased expenditure on NPD projects is no guaranteed of increased NPD success.
Rationale: Many electronics firms are not monitoring their NPD costs. They have no way of measuring how much the development of a new product costs, they have no visibility of which NPD projects are over budget or of how long it takes new products to breakeven. Until firms can account for where they are spending NPD resources it is possible that extra resources will simply allow developers to spend extra time on projects but will not improve the chances of commercial success. Support for this hypothesis comes from the survey results – R&D expenditure as a percentage of sales was not linked to new product success in any groups of firms surveyed. In other words, firms that spend more developing new products are not necessarily more successful. Within the case studies it was also seen that, with the exception of one firm, there was limited awareness of the costs of developing a new product and no evidence that NPD costs were being controlled. The literature reviewed suggests that a lack of financial resources is a barrier to product innovation (Fitzgerald & Breathnach, 1994; Roper et al. 1996) – this can be interpreted as showing that firms do not develop new products because of a lack of finance. However, this study has included only firms who are already developing new products and it appears that within this group increasing NPD expenditure is no guarantee of increased new product success.

H3: Reducing time-to-market in the electronics industry is only linked with increased new product success under specific conditions.
Rationale: The importance of time-to-market appears to depend on the sector within which a firm is operating. In some sectors, where technology is changing rapidly, time-to-market is a critical factor and only firms who can launch new products within the time frame required can survive. In other sectors time-to-market is not such a critical issue and launching the 'right' product is more important that launching a product at the 'right' time. Support for this comes from the INTERPROD survey that showed that NPD cycle time is only linked with new product success in large UK firms. In large Irish and small UK firms cycle time correlated positively with new product success suggesting that products with longer cycle time were more likely to be successful. The case studies also suggested that the importance of cycle time depended on the sector within which firms operated. Time-to-market was an important issue for two of the firms studied but for the third firm product reliability and functionality appeared to be more important than the launch date. Much of the literature examined on cycle-time practices assumed that delivering products on time was critical to all firms and concentrated on techniques to reduce NPD cycle-times. Frequently much older references (Vesey, 1991) were quoted to show the importance of releasing new products on time – a study of the importance of time-to-market across different industry sectors could help to clarify this issue.

H4: Increased use of project management tools and techniques in NPD projects will increase the success of NPD in Irish electronics firms.
Rationale: The Irish firms stood out from the others in the study based on their limited use of project management techniques. This hypothesis is supported by both the survey and the case studies. Analysis of the Irish INTERPROD data showed that there is no correlation between project management skills and new product success. Further investigation of this finding in the case studies revealed that Irish electronics firms are not using project management. The largest of the three firms studied was beginning to consider using project management techniques at the time of the case study. The two small firms were not using project management. The literature reviewed suggests that the use of strong and competent project managers is linked with success (Wheelwright & Clark, 1992; Norrgren & Schaller, 1999), this is an areas where Irish firms need to improve.

H5: The 'best' method of organising NPD projects is dependent on a firm's size, technology and market.
Rationale: An organisational structure that is suitable for a large firm developing complex products will not necessarily work for a small firm using a simpler technology. The analysis of the INTERPROD data showed that there was no link between the method used to organise a NPD project and new product success. This is supported by the case studies where it is clear that the most suitable way of organising NPD projects depends on the size of the firm and also the type of projects undertaken. The literature also suggests that different organisational structures are suitable for different sized firms. Studies of large firms recommend heavyweight project teams (Cooper, 1994; Clark & Wheelwright, 1992) while studies of small firms show that they benefit from 'organisational flexibility' (Rothwell & Dodgson, 1991).

H6: The NPD process stages that are linked with success in small firms are different from those linked with success in large firms.
Rationale: Large electronics firms tend to manufacture product in large volumes therefore introducing a product to manufacturing is a critical process stage. The Irish INTERPROD data shows that manufacturing and marketing start-up are critical activities for large firms. Small firms are often younger and need to concentrate on developing high quality products and also on developing markets for their products. The INTERPROD data shows that it is important for small firms to be proficient at prototype development and market development. Within the literature different studies have also stressed the importance of different aspects of the NPD process (Cooper & Kleinschmidt, 1986, 1987; Song & Montoya-Weiss, 1998; Holger 2002). The evidence suggests that the relative importance of the different stages of the development is contingent on firm size, technology and market.

H7: Marketing synergy is more important for new product success in Irish electronics firms than technical synergy.
Rationale: The Irish INTERPROD survey found that marketing synergy is linked with new product success but technical synergy is not. The analysis suggests that it is easier for Irish electronics firms to develop new technologies than to develop new markets. This is also seen in the new products that Irish electronics firms develop. Of the 63 projects included in the survey only 20% were developed under high market uncertainty while 32% were developed under high technical uncertainty. Many earlier studies support a link between marketing and technological synergy and

new product success, however in this study no link between technological synergy and success was found. This hypothesis is linked to and supported by H12 below that proposes that marketing skills are more important than technical skills.

H8: Increased R&D/Marketing integration is most strongly linked with increased new product success in firms with between 20 and 100 employees.
Rationale: The INTERPROD survey shows that R&D/marketing integration is linked with new product success in firms with between 20 and 100 employees. Very small firms often operate like teams with good communications between different functional groups on all projects, R&D/Marketing integration is not a success factor; eFlex was an example of this. Large firms usually have processes in place that require cross-functional involvement in the development of all new products, again R&D/Marketing integration is not a success factor – Telephase is an example of this. However, firms that fall between these two groups can have difficulties in managing R&D/Marketing integration. The literature supports the link between R&D/Marketing integration and new product success (Griffin & Hauser, 1996; Song & Parry, 1997; Leenders & Wierenga, 2002) but does not identify the impact of firm size as highlighted by this study.

H9: Increased top management support is most strongly linked with increased new product success in firms with between 20 and 100 employees.
Rationale: The INTERPROD survey only found a link between top management support and new product success in firms with between 20 and 100 employees. The case studies offer an explanation for this by showing that top management support is normally guaranteed for NPD projects in very small and large firms. However, firms with between 20 and 100 employees have difficulty in ensuring that top management are supportive of all NPD projects. The rationale for this is similar to that offered for H8. The literature supports a link between top management support and new product success (Cooper & Kleinschmidt, 1995; Swink, 2000) but does not explore the impact of firm size.

H10: Product compatibility is the new product characteristic that is most closely linked with new product success in small firms.
Rationale: Small firms have difficulty in entering markets with radical new products; customers are unlikely to take the risk involved in buying

unknown products from unknown firms. The INTERPROD survey showed that using product compatibility as a positive selling point was linked with new product success in small but not large Irish firms. The case of eFlex also supports this hypothesis. As a small firm, eFlex had difficulty selling a product based on a very novel technology. They were more successful when they sold their technology embedded in the products of larger firms. There is support for this in the literature where small firms are advised to avoid highly diversified or differentiated new products (Hoffman et al. 1998; Yap & Souder, 1994; Meyer & Roberts, 1986).

H11: Small firms are most successful with innovative new products that address specific or niche customer needs.
Rationale: Small firms are most successful when operating in market niches and developing innovative new products to satisfy specific customer needs. In this way small firms avoid direct competition with large firms. The INTERPROD survey results show that small Irish firms have most success when they develop new products that address the problems that customers have with existing products and when customers have a great need for the product. The case studies also provide some support for this hypothesis; the two small firms developed products for specific customers while the large firm had a much wider customer base. Within the literature the idea that small firms tend to be more successful in market niches is also generally accepted (O'Gorman, 2000; Hoffman et al. 1998).

H12: In small firms, the availability and level of marketing skills is linked with new product success but there is no link between technical skills and success.
Rationale: It would appear that most small firms either have all the necessary technical skills or can acquire them easily while the availability and level of marketing skills differentiates between success and failure. The survey results show that there is a strong correlation between marketing skills and new product success in small firms but that technical skills are not significantly linked with success. Earlier INTERPROD publications also support this finding (Souder et al., 1997).

These twelve hypotheses provide an outline for the factors that have an impact on the success of new products in small Irish electronics firms. The survey and case study have also resulted in insights into the management of NPD in Irish electronics firms outside the frame of the research model. The overall findings of this research are summarised in Section 10.5 below.

10.5 Key Survey and Case Study Findings

The research question posed at the start of this study is repeated below.

How do small Irish electronics firms manage the development of new products, and are the success factors that apply to small Irish firms different from those that apply to large Irish firms?

The research model described in Section 10.4 has addressed this question by identifying within the literature the factors that have been found to have an impact on NPD success and explaining how these factors apply to small and large Irish electronics firms. However, the key findings of this research can also be summarised in terms of their usefulness to managers, policy makers and future researchers.

10.5.1 Implications for Managers of Small Firms

The research has provided some points of advice for managers in small firms who have responsibility for the development of new products.

1. *NPD costs should be closely monitored and controlled.* NPD activity is often measured by the amount that a firm invests in developing new products. But NPD expenditure tells nothing about how effective a firm is at product development or about what value a firm gets for its investment in NPD. NPD is an expensive process; it is very easy to waste a considerable amount of money in developing new products. Managers should set budgets for NPD projects, and be able to monitor actual expenditure against these budgets. Without a basic level of financial control managers cannot make informed go/no-go decisions or determine which new products have a positive rate of return and which do not. Increasing NPD expenditure is no guarantee of increased NPD success.
2. *NPD processes and structures that work in large firms do not necessarily work in small firms.* Small firms should structure their NPD projects in a way that suits their business and size; they should not simply try to emulate larger firms. Larger firms tend to have more formal NPD procedures but in small firms there is no significant link between process formality and new product

success. Small firms should not attempt to implement a very formal NPD process unless it is necessary – it will be expensive in terms of administration and will not necessarily improve success.

3. *Involving external organisations in NPD projects can be beneficial but it is difficult to manage effectively.* Small firms are frequently advised to network and involve external organisations in their NPD projects. However, not all small firms have the ability to effectively manage this process and they can find themselves spending a lot of money and not getting the product/technology they require within a reasonable timeframe. Before small firms involve external organisations in their NPD projects they should ensure that they have the skills to effectively manage the process.
4. *Small firms are most successful with new products that are aligned with their existing markets and products; are compatible with similar available products and address niche markets and specific customer needs.* It is difficult for small firms to be successful with radical new products. Small firms are most successful when the products they develop take advantage of their existing skills – particularly their marketing skills. Additionally, small firms are more successful when they develop products that are compatible with existing products and technologies. Finally, larger firms often ignore smaller niche markets and specific customer needs but these can offer good sources of business for small firms.
5. *As firms grow and lose some of the benefits of good communications and shared vision often associated with very small firms, they need to pay more attention to the way that NPD projects are managed.* Very small firms and very large firms do not have a problem with R&D/Market integration and top management support on NPD projects. However, in firms with between 20 and 100 employees these factors need to be carefully managed to ensure new product success. As small firms grow beyond 15-20 employees they need to re-assess the way in which they manage NPD projects.

10.5.2 Implications for Policy Makers

In addition to the advice for managers of NPD projects in small firms, the following could benefit policy makers who are responsible for supporting innovation and product development in small firms.

1. *Increased funding for NPD projects is not liked with increased success. Small Irish electronics firms benefit more from improved management and marketing skills than from additional finance for NPD.* Increasing NPD expenditure is no guarantee of success, R&D grants are probably not effective – firms need to improve the way that they are managing NPD projects rather than spend more on NPD projects. Funds could be better deployed in improving the management and marketing skills of small firms, for example in assisting firms to develop marketing skills, implement project management techniques, or control their NPD expenditure i.e. account for where the money goes.
2. *Small Irish electronics firms need assistance in addressing new markets.* Irish firms appear to be reluctant to address new markets and do not seem to have the appropriate marketing skills to develop new products for new markets. Small firms are more successful with new products that are launched into their existing markets but there is also evidence that they are reluctant to explore new markets possible due to a lack of marketing skills. In order to grow and increase employment, most small firms will need to expand their customer base – small firms need assistance in doing this.
3. *With a few exceptions, large Irish electronics firms are involved mainly in incremental product development.* Much of the NPD happening in large firms is incremental. This limits the ability of this sector to generate additional employment through links back into the indigenous sector and the generation of spin off firms. Much funding has been provided to this sector to support its R&D activity, the effectiveness of this funding should be reviewed. Also the lack of a serious R&D function in many of the large electronics firms located in Ireland suggests that they can easily relocate out of this country.
4. *Small firms with limited resources and management skills are developing innovative new products.* Small firms should be supported in their product innovation efforts. They are developing more innovative new products than many of the large firms despite the fact that they are very limited in the resource that they have available for NPD. They should be supported to

develop the skills (marketing, project management, financial control) that they need to improve their NPD success

10.5.3 Implications for Future Researchers

Details of directions for future research are described in Chapter 11, but at a general level the following two points emerge from the research.

1. *NPD success factors vary with firm size, location, and market environment, researcher should avoid proposing one-size-fits-all solutions.* Researchers seeking to identify success factors in NPD should be aware that there is no one solution. What works in one situation or for firms of a particular size will not always work in other situation. In particular, results that are based on the management of NPD in large firms should not be assumed to apply to small firms.
2. *Some of the basic assumptions surrounding research into the management of NPD that may have applied several years ago need to be reconfirmed.* It is worth reinvestigating some of the basic assumptions e.g. that time to market is always a key variable, the interaction with external organisation is always good, that top management support and R&D/Market integration are linked with success in all firms. Over the past 10 – 20 years, markets have changed, NPD management has changed, and technology has changed. Basis assumptions that have guided many researchers in the '80s and '90s may not still apply and should be re-examined.

10.6 Conclusion

This chapter contained an analysis of the results of the three case studies presented in Chapter 9. The analysis focused on five areas that were identified based on the results of the INTERPROD survey. These were cycle-time practices, resources for NPD, methods of organising NPD, involvement of external organisations, and project management skills. The discussion of these topic areas linked the case study results back to the results of the INTERPROD survey; the case studies were used to further describe and to explain some of the results from the survey.

The results of the case studies were then used to update the research model and to develop a set of hypotheses based on the relationships shown in the model. These hypotheses help to structure future research into the management of NPD at small firms by highlighting possible relationships and explanations for some of these relationships.

Finally, this chapter presented a summary of the key findings of the survey and case studies in the form of advice for managers, policy makers and future researchers. The final chapter in this thesis will summarise the overall research carried out and propose directions for future research.

CHAPTER ELEVEN

CONCLUSION

11.1 Introduction

This chapter contains the conclusions to this research project. It is divided into four sections. The first summarises the thesis showing the sequence of events from the development of the research question, through the review of literature to support the study, the definition of a research method, the implementation of the research plan and finally the results that emerged. The second section identifies the way in which this study has contributed to knowledge about the management of NPD in small firms; it also discusses some of the limitations of this research. The third section proposes several directions for future research by highlighting issues that have emerged but have not been fully explored. The conclusion of this chapter is in the final section.

11.2 Summary

This thesis began with the identification in Chapter 1 of New Product Development as a critical factor in increasing competitiveness both at a company and a national level. The importance of the electronics industry and of small firms to the Irish economy was demonstrated and several issues emerged about the way in which small Irish electronics firms develop new products. These issues were discussed and summarised in the research question:

How do small Irish electronics firms manage the development of new products, and are the success factors that apply to small Irish firms different from those that apply to large Irish firms?

Chapter 2 reviewed new product development success factors highlighting references to small firms. The review was structured around three topic areas 1) Strategic & Market Factors, 2) Development Process Factors and 3) Organisational Factors, several factors within each area were addressed.

The literature suggested that there are differences between the way in which NPD is managed in small and large firms and also that different success factors apply to large and small firms. Within the literature review one particular stream of publications, based on the INTERPROD study, proved to be comprehensive in its analysis of the factors that distinguish between new product success and failure. Therefore it was decided to examine the INTERPROD study in more detail. Chapter 3 reviewed the 13 publications that had resulted from the INTERPROD project. Several general themes that emerged from the publications were summarised. Findings that related to the management of small firms or to firms in small economies similar to Ireland were noted.

The model that guided the research was presented in Chapter 4. The model was broadly based on the INTRPROD project but was modified to include firm size as a variable that affects the manner in which NPD is managed and also the determinants of new product success. The research model, see Figure 11.1, suggested that firm size and location have an impact on the management of NPD at a project and company level, and also on the way in which NPD management practices are linked with success.

Based on the relevance of INTERPROD to the research question it was decided to adopt a multi-method approach. The research strategy chosen was to use a quantitative survey, INTERPROD, to identify some of the differences between NPD management in small and large firms and then to use qualitative case studies to explore these differences in more detail, the research method was presented in Chapter 5.

The research model proposed in Chapter 4 was developed throughout this thesis; the four stages of the model are presented in Figure 11.1. This model was tested initially by replicating the analysis described in previous INTERPROD publications on the Irish INTERPROD data. This analysis confirmed that firm size and location influence the way in which NPD is managed and also identified success factors that apply at a project level. An updated model was presented in Chapter 6. No evidence emerged about company level innovation management practices.

The next stage in testing the research model and gaining a deeper understanding of NPD management in small Irish electronics firms was an in-depth analysis of the Irish and UK INTEPROD data this was presented in Chapter 7. The discussion presented in Chapter 8 supports the proposal

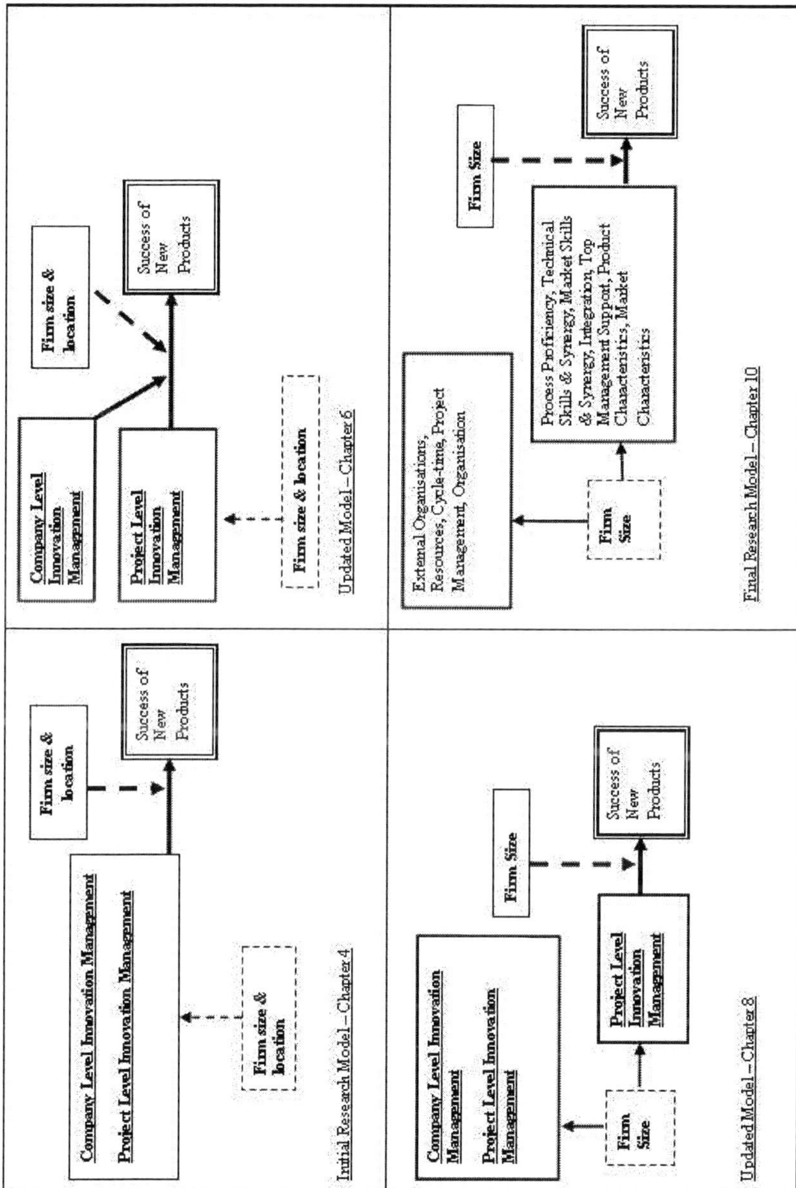

Figure 11.1 Research Model Evolution

that firm size and location impact the management of NPD and the success factors that apply. The updated model presented at the end of Chapter 8 focuses on the Irish situation therefore the location variable was dropped and the findings from Chapter 6 were generally supported. All but two of the project level management practices were linked with new product success and were affected by firm size, but while the company level management practices were impacted by firm size they were not linked with new product success.

The final stage in this study was to use three case studies to explore in more detail the findings of the model presented in Chapter 8. The case studies, presented in Chapter 9, focused on the five variables that were not linked with new product success and were useful in explaining why no links were found. The findings of the case studies were used to propose the final research model in Chapter 10 and to develop a set of hypothesis about the management of NPD in small firms. These hypotheses help to structure future research into the management of NPD at small firms by highlighting possible relationships and explanations for some of these relationships.

11.3 Contribution to Knowledge

This study sought to gain a better understanding of how small firms develop new products and also to identify which management practices were linked with new product success in small Irish electronics firms. Because of the extensive knowledge available about the management of NPD in large firms the approach taken was to determine whether the NPD success factors that applied to large firms also applied to small firms. The data collected and analysed in this thesis have resulted in the following contributions to the body of knowledge about the management of NPD.

1. The management of NPD in small firms is not the same as that in large firms. This study has shown that success factors and management practices that are found in large firms do not necessarily apply to small firms. This has implications for managers, researchers and policy makers. Specific differences between small and large firms have been identified in the following areas:
    a. NPD resources and expenditure,
    b. Organising for NPD,
    c. NPD process proficiency,
    d. Marketing & technical skills and proficiency,

   e. R&D/marketing integration and top management support,
   f. New product and market characteristics.
2. Data were collected about the management of NPD in Irish firms. The INTERPROD study of NPD success and failure was conducted in 15 different countries before being carried out in Ireland. This is has provided a unique insight into the management of NPD in Irish firms and has allowed NPD management in Ireland to be compared with that in other countries. The key features of NPD management in Ireland that were highlighted by this study were:
   a. Large Irish electronics firms are involved mainly in incremental product development.
   b. Irish firms are good at developing technically difficult new product but are slow to enter new markets.
   c. Irish electronics firms do not have well developed project management skills.
3. The study also resulted in two findings about the management of NPD that are not specific to small or Irish firms.
   a. Involving external organisations in NPD projects is not a simple process nor is it a guarantee of new product success.
   b. The importance of time to market is dependent on the environment in which a firm operates. It is not always a good strategy to speed up the NPD process.
4. A model for the management of NPD in small firms was developed. The model will be useful for managers and policy makers identifying NPD success factors and highlighting the differences between NPD management in small and large firms.

11.3.1 Research Limitations

This study has provided many interesting findings however there are ways in which it could have been improved, these are summarised below.

1. The advantage of using INTERPROD was that it allowed comparisons to be made between NPD management in Ireland and other countries. However, there were some drawbacks associated with using a set of questionnaires that had not been developed specifically for this study. Some of the areas addressed by INTERPROD were not of particular interest for this study and there were topics (e.g. NPD resources) that could have been examined in more detail. If this study were to be repeated

the INTERPROD survey could be tailored more to suit the Irish situation, however it should be noted that the implication of doing this would be a loss of comparability with other countries.

2. The Irish INTERPROD data were compared in detail with the UK INTEPROD data. While this comparison was useful in placing the Irish results in context, in hindsight it may have been useful to compare the Irish data with those from other countries that are more similar economically to Ireland for example, Finland, New Zealand or Singapore.
3. The three case studies were selected to reflect the structure of the INTERPROD sample, but they were three very different firms. While this was useful in examining the diverse situations within which firms operate and the different ways that they manage the development of their new products it would also have been interesting to examine a set of similar small firms. This would have resulted in a richer understanding of the issues that apply to small firms. Additionally, extending the case study research to include more firms would also have strengthened the explanations provided.
4. There was a time delay between the collection of the INTERPROD and the case studies data. All three firms changed between responding to INTERPEROD and being involved in the case studies. At one level this weakens the 'chain of evidence' between the cases and survey however it also reflects the reality of conducting this type of research where the researcher has no control over the sources of information.

11.4 Future Research

The objective of this research project has been to develop a theory about how the management of NPD in small firms is different from that in large firms. This has resulted in the development of a research model and a set of twelve hypotheses. Clearly the next stage is to test these hypotheses, though this is beyond the scope of this study. Several of the topics covered by the hypotheses presented in Chapter 10 are complex and extensive and could form the basis for the development of additional models and research projects. Some of these research topics that appear particularly interesting are summarised below. For each research topic a rationale, research objectives and a proposed research method will be outlined.

1. *Involvement of External Organisations in NPD Projects.*

   Rationale: The INTERPROD survey did not find a link between the involvement of external organisations in NPD projects and increased new product success. The fact that these results were consistent in Irish and UK firms makes them more robust and worthy of further investigation. This was an unexpected result especially considering the fact that development agencies tend to encourage small firms to interaction with external organisations when developing new products. The case studies offered some explanation for the lack of a relationship between NPD success and the involvement of external organisations in NPD projects. However given the general assumption that networking increases a firm's innovation output this issue deserves further investigation.

   Objectives: To develop a framework for the involvement of external organisations in NPD projects – the framework would address when external organisations should be involved, how the relationship should be managed and what outcomes to expect. A secondary objective would be to quantify the extent and nature of the involvement of a sample of firms with external organisations in NPD projects and to determine the relationship between such involvement and new product success.

   Methodology: It is likely that this project would gain the support of the development agencies here in Ireland or even the Industrial Liaison Office within a university. If this were the case an appropriate research strategy would be to conduct a survey within a particular industrial sector (or a group of sectors) to determine the extent and nature of external involvement in NPD projects and also to examine the link with new product success. A series of case studies would also need to be conducted to examine how relationships with external organisations are managed, this could possible be set up to examine several successful and unsuccessful cases.

2. *Marketing Strategies for Small Technology Based Firms.*

   Rationale: The results of the INTERPROD survey suggested that small firms are more successful in mature markets and large firms are more successful in quickly growing markets – this was consistent in UK and Irish firms. INTERPROD also showed that Irish firms are reluctant to enter new markets – yet in most cases they need to do so in order to grow – and that small Irish firms are most successful with new products that are aimed at their existing markets.

Objectives: To identify successful market growth strategies for small technology based firms across a range of sectors.

Methodology: This study should examine several sectors – it is likely that different growth strategies will apply in different sectors. A series of surveys examining issues such as market strategies, distribution channels, marketing diversity, technical complexity, new product success, market growth and intensity, and barriers to addressing new markets; would be useful in achieving this research objective.

3. *Room for Improvement in Ireland's Science, Technology and Innovation Policy*

Rationale: The INTEPROD survey highlighted several issues that have implications for policy makers; increased funding is not necessarily linked with increased NPD success, small firms need assistance in addressing new markets, small firms would benefit more from increased skills than from increased funding, large firms are involved mainly in incremental NPD. Previous research by this author (Ledwith, 1997) also shows that large firms receive significantly more R&D grant assistance per employee than small firms. All of this suggests that the way in which innovation is funded in Ireland should be reviewed.

Objectives: To evaluate existing support for new product development in the technology sector in Ireland and identify how it can be changed to increase new product output and success in Irish firms.

Methodology: The cooperation of the development agencies would be important for the successful completion of this project. The first stage would be to examine historical data relating to supports for innovation, how they have been allocated and what they have achieved. The second stage would require a survey to examine which supports firms have received whether they have improved new product success, and to explore the needs of firms in developing new products and how these needs might be addressed.

4. *Project Management in New Product Development*

   Rationale: The INTERPROD survey found that most Irish firms are not using project management techniques to manage NPD projects but that those that do experience a reduction in NPD cycle times. Additionally, the case studies showed that it is not straightforward to implement project management techniques in NPD projects yet firms are in need of some method of controlling NPD cost, time and performance.

   Objectives: To develop a project management methodology that is suitable for managing NPD projects in technology based firms.

   Methodology: This research project would involved working closely with a group of firms to develop a set of project management techniques that would be tailored to suit NPD projects. The most suitable research method would be action research.

The above list of projects is by no means exhaustive, for example the issues of controlling NPD costs or time to market could be explored as part of project 4 above but they could also be explored as individual projects. The projects defined above represent well defined areas for future research that build on the results of this study.

11.5 Conclusion

This final chapter has provided a summary of the thesis, from the development of the research question to the final interpretation of the results. It has summarised the contribution that this study has made to knowledge about the development of new products by identifying the differences between NPD in small and large firms, providing unique data about NPD development in Ireland and developing a model for NPD in small firms. The limitations of the study were discussed and finally several areas for future research were described.